Jaypee Gold Standard Mini Atlas Series®

EMBRYOLOGY

Jaypee Gold Standard Mini Atlas Series®

EMBRYOLOGY

Second Edition

Roopa Kulkarni
Senior Professor and Former Head
Department of Anatomy
MS Ramaiah Medical College
Bengaluru, Karnataka, India

Varsha S Mokhasi
Professor and Head
Department of Anatomy
Vydehi Institute of Medical Sciences
and Research Center
Bengaluru, Karnataka, India

Foreword
HR Krishna Rao

The Health Sciences Publisher

New Delhi | London | Philadelphia | Panama

Jaypee Brothers Medical Publishers (P) Ltd.

Headquarters
Jaypee Brothers Medical Publishers (P) Ltd.
4838/24, Ansari Road, Daryaganj
New Delhi 110 002, India
Phone: +91-11-43574357
Fax: +91-11-43574314
E-mail: jaypee@jaypeebrothers.com

Overseas Offices

J.P. Medical Ltd.
83, Victoria Street, London
SW1H 0HW (UK)
Phone: +44-20 3170 8910
Fax: +44(0) 20 3008 6180
E-mail: info@jpmedpub.com

Jaypee-Highlights Medical Publishers Inc.
City of Knowledge, Bld. 237, Clayton
Panama City, Panama
Phone: +1 507-301-0496
Fax: +1 507-301-0499
E-mail: cservice@jphmedical.com

Jaypee Medical Inc.
325 Chestnut Street
Suite 412, Philadelphia, PA 19106, USA
Phone: +1 267-519-9789
Email: jpmed.us@gmail.com

Jaypee Brothers Medical Publishers (P) Ltd.
17/1-B, Babar Road, Block-B, Shaymali
Mohammadpur, Dhaka-1207
Bangladesh
Mobile: +08801912003485
E-mail: jaypeedhaka@gmail.com

Jaypee Brothers Medical Publishers (P) Ltd.
Bhotahity, Kathmandu, Nepal
Phone: +977-9741283608
E-mail: kathmandu@jaypeebrothers.com

Website: www.jaypeebrothers.com
Website: www.jaypeedigital.com

Inquiries for bulk sales may be solicited at: jaypee@jaypeebrothers.com

Jaypee Gold Standard Mini Atlas Series®: Embryology

First Edition: 2010

Second Edition: **2016**

ISBN: 978-93-5250-159-5

Printed at Sanat Printers

Dedicated to

Our beloved students

PES Institute of Medical Sciences & Research
KUPPAM - 517 425, CHITTOOR DISTRICT, ANDHRA PRADESH
Tel : 08570-202299, 256736, 207777, 203366 Fax : 08570-257261, 257200
Email : principal@pesimsr.pes.edu www.pesimsr.pes.edu

Dr. H.R. KRISHNA RAO MBBS., MS (Anatomy)
Principal & Prof. of Anatomy

Foreword

Embryology is an integral part of curriculum in anatomy for MBBS, MD—Anatomy, Postgraduate, Dental and allied courses.

Rarely a main question is asked in the undergraduate examinations, but it is of great significance in the PG entrance examination all over the country, and there is a separate paper in PG Theory Examination in Anatomy.

Due to 1-year program for the 1st year MBBS, and at times, curtailed, due to many reasons, reading major textbooks of embryology providing vast knowledge is time consuming, and may not be necessary at undergraduate level.

Hence, *Jaypee Gold Standard Mini Atlas Series: Embryology*, by Dr Roopa Kulkarni and Dr Varsha S Mokhasi is a perfect book, similar to a coffee table book, giving a rapid to-the-point revision of the subject, for preparation before any examination and suffice for coming out with excellent performance in the assessment, or in examination.

The effort by both of them is very thorough, which shows their wisdom, teaching experience and knowledge of what is expected from the students, for assessing.

I am sure, the book will be extremely popular and express my best wishes for the same.

HR Krishna Rao MBBS MS (Anatomy)
Principal and Professor of Anatomy
PES Institute of Medical Sciences and Research
Kuppam, Chittoor, Andhra Pradesh

Preface to the Second Edition

Embryology is a section of anatomy, which forms a basis to understand the gross anatomy. It forms an important part to understand many clinical aspects in pediatrics and pediatric surgery, especially in dealing with congenital anomalies. Though there are plenty of photographs to explain the embryology, most of them have to be imagined to understand various stages in the development of the human body.

The development and growth of the human body is amazing and very useful in understanding the clinical anatomy.

In this regard, we have made an effort to make the subject as simple as possible to understand the basics in embryology by using the photographs of plastic and clay models used to study embryology. In this edition, though the description has been kept very short, some of the detailed description have been added to enhance the knowledge in basics of embryology and also kindle the interest in exploring more about the development of the human body and aimed at the undergraduate medical students to make the subject simple, easy to understand and reproduce.

Roopa Kulkarni
Varsha S Mokhasi

Preface to the First Edition

The subject *Human Embryology* is the basis for understanding the human anatomy. It is the most essential subdivision, which helps to understand various clinical subjects. Therefore, the content of the subject cannot be reduced. The hurdle in understanding embryology is to imagine the developmental stages. Moreover, the subject is vast and the duration for learning normal Human Embryology by medical students in the first phase is reduced from one and half years to one year. Therefore, it was thought to make the subject Human Embryology simple, easily understandable and reproducible with the help of the models, which give three-dimensional visualization.

In this regard, we express our deep sense of gratitude to all those who have inspired and created interest in Human Embryology and the modelers who have taken pains to bring the structures to near reality with proportionate magnification of various parts of the embryo and the fetus.

The development of human body is so fascinating that the efforts that are made in bringing out this book will not only help in understanding and performance in the examination but also kindle the interest in exploring more about the development of the human body.

Roopa Kulkarni
Varsha S Mokhasi
Shailaja Shetty

Acknowledgments

It gives us immense pleasure to express our gratitude to Dr MR Jayaram, Chairman, Gokula Education Foundation, for giving an opportunity to be a part of the organization and carry out this work.

We thank Dr GV Guruprasad, Chief Executive Officer, GEF (Medical), for the constant encouragement in academic activities of the institution.

We also express our deep gratitude to Dr AC Ashok, Principal and Dean, MS Ramaiah Medical College and Hospitals, for his utmost concern and support for students.

We thank Mr Chetan, who has helped us in the modification and digitalization of the photographs.

We are greatly indebted to the unnamed modelers in Anatomy, without whom the book would not have reached the present stage. The imagination and communication in the field of Embryology would have become a dream for most of the teachers and students without the models prepared by them.

We thank the staff members in the department of anatomy for their constant support, constructive suggestion and supporting the use of the book by undergraduates.

We are extremely thankful to Shri Jitendar P Vij (Group Chairman), Mr Ankit Vij (Group President), Mr Tarun Duneja (Director-Publishing) and their team of M/s Jaypee Brothers Medical Publishers (P) Ltd, New Delhi, India, for their effort in bringing out our ideas in print form.

We thank all undergraduate and postgraduate medical students who suggested and encouraged us to go in for the second edition.

We thank one and all.

Contents

Human Embryology

Chapter 1

INTRODUCTION

Human embryology is that branch of Anatomy, which deals with the development of the individual.

The term Embryology, appropriately called developmental anatomy, is applied to the various changes, which take place during the growth of an animal from the egg to the adult condition.

Embryology may be studied from two aspects:

1. That of ontogeny, which deals only with the development of the individual.
2. That of phylogeny, which concerns itself with the evolutionary history of the animal kingdom.

TERMS USED IN EMBRYOLOGY

Prenatal Period

- Oocyte
- Sperm
- Zygote
- Fertilization age—1st day
- Cleavage

- *Morula:* 12 to 32 blastomeres—3 to 4 days
- *Blastocyst:* 5th day
- *Implantation:* 6th day
- *Gastrula:* Formation of trilaminar disc—up to 3rd weeks
- *Neurula:* Formation of neural tube—3rd to 4th weeks
- *Embryo:* Stage extends up to end of 8th weeks—Formation of all major structures
- *Conceptus:* Includes embryo and fetal membranes
- *Fetus:* Unborn offspring
- *Abortion:* Premature stopage of development and expulsion of conceptus
- *Trimester:* A period of 3 calendar months during pregnancy (I, II, III trimesters).

Ist trimester is the critical stage of development.

Postnatal Period

- *Neonate:* First 1 month after birth
- *Infant:* 1st year after birth
- *Childhood:* 13 months to puberty
- *Adolescent:* 11 years to 19 years
- *Adulthood:* 18 years to 21 years
- *Adults:* Up to 40 years
- *Middle age:* 41 years to 55 years
- *Old age:* Above 55 years.

From fertilization to birth—approx 280 days are required—equivalent to 40 weeks/10 lunar months.

Age of an embryo or fetus is expressed in terms of weeks or lunar months of intrauterine life (IUL).

SPERMATOGENESIS

It is a term used to describe the formation of spermatozoa from spermatogonia **(Fig. 1.1)** of 22 days.

It takes place in the seminiferous tubules of the testes.

The wall of the seminiferous tubule has a basement membrane on which the germ cells are situated. These germ cells are supported by columnar cells.

The germ cells are called the **spermatogonia** and the columnar supporting cells are called **Sertoli cells**.

The gamatogenic cells at the basement membrane are called **spermatogonia**. They are of two types and they are **type 'A'** and **type 'B'**.

Type 'A 'cells are the stem cells which divide again into type 'A' and 'B' cells.

Type 'B' cells divide by **mitosis** into **primary spermatocytes**. These cells are the largest among all the spermatogenic cells and contain darkly stained nucleus. Primary spermatocytes have prolonged prophase.

Each primary spermatocyte divides by **first stage of meiosis** into **secondary spermatocyte**. These cells are smaller. Each secondary spermatocyte contains half the number of chromosomes (22 autosomes and one sex chromosome). Conversion of primary spermatocyte to secondary spermatocyte requires about 8 days of time.

The secondary spermatocytes now divide **by second stage of meiosis** into **spermatids**. This requires about 16 days.

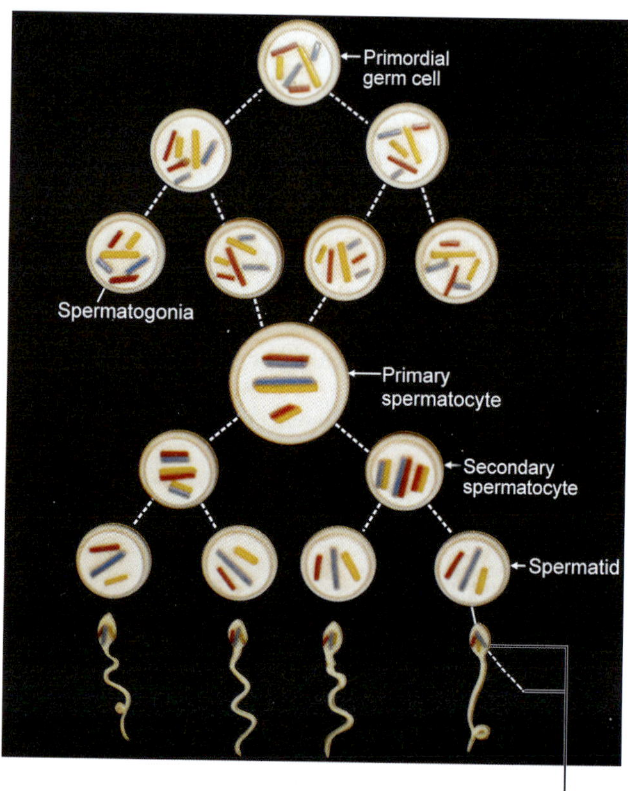

Fig. 1.1: Spermatogenesis

The spermatids undergo **metamorphosis** and become **spermatozoa** in about 24 days.

The conversion of spermatids to spermatozoa is called **spermiogenesis**.

The total duration of spermatogenesis is 64 days. Maturation of the sperms takes place in the epididymis and vas deferens.

The spermatogenesis is controlled by follicle stimulating hormone (FSH), luteinizing hormone (LH) and interstitial cell stimulating hormone (ICSH) of anterior pituitary and adrenal gland. These hormones stimulate Leydig cells of testis to secrete testosterone. The mature sperm is about 60 micrometers in length and has head, body and tail.

Spermiogenesis

In this process, the spermatids undergo metamorphosis and become spermatozoa.

The changes which take place in this stage are:

- The **Golgi apparatus** becomes the **acrosomal cap**,
- The **nucleus** forms the **head of the sperm**,
- **Mitochondria** contribute to body or the middle **piece**,
- The centrosomes split and form proximal and distal centrioles that help in the formation of middle piece
- **Rest** of the **cytoplasmic contents—the microtubules, microfilaments—form** the **body and the tail** of the spermatozoon.

The remaining cytoplasm form residual body, which is extruded out of spermatozoon. Acrosomal cap contains hyaluronidase which dissolves the cumulus oophorus cells and helps the head of the sperm to penetrate through the zona pellucida.

The tail acts as flagellum and propels the sperm forward. The sperms are attached to the apices of the Sertoli cells or inserted into the notches present in the apices of Sertoli cells till they mature/become motile and are released into the lumen of the seminiferous tubules.

The sertoli cells are the supporting cells, which also nourish, phagocytose the waste products and help in maturation of the sperms. After the sperms mature, they become motile.

About 200 to 300 million of sperms are present in 2 to 5 mL semen (in one ejaculation). The male in whom the sperm count is less than 20 million is considered as sterile.

OOGENESIS

The development and maturation of oocytes in the ovaries is called Oogenesis **(Fig. 1.2)**. Ovaries are paired organs situated in the pelvis. Each ovary is an almond shaped structure and consists of an outer cortex and an inner medulla.

The cortex contains the ovarian follicles at different stages of maturation.

To begin with there are oogonia in the early fetal life. These oogonia are not surrounded by follicular cells therefore are naked oogonia. There are about 6 million oogonia by 5th month of intra-uterine life. At the time of birth, only 2 million remain and rest of them disintegrate and disappear. At the end of the fetal life, all the oogonia are converted into primary oocytes which are surrounded by a single layer of squamous or low cuboidal follicular cells. Only 40,000 oocytes remain in each ovary of which only 400 undergo full maturation during the reproductive life of a female.

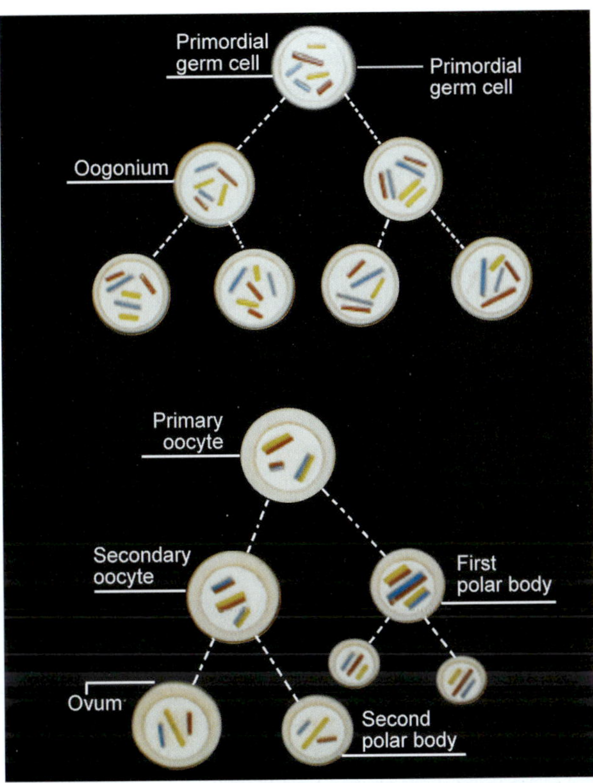

Fig. 1.2: Oogenesis

The mass of oocyte and the surrounding follicular cells is called the ovarian follicle. The primary oocyte and single layer of squamous follicular cells is called **primordial ovarian follicle**. The **primary oocytes** are surrounded by a layer of cuboidal follicular cells. Under the influence of follicle stimulating hormone the squamous follicular cells gradually become cuboidal, columnar and multilayered.

The oocyte in the primary follicle divides into **secondary oocyte** and a **first polar body** by **meiosis I**. The follicular cells simultaneously divide by mitosis into many cells which form multiple layers of cells around the secondary oocyte and the 1st polar body. Thus, **secondary follicle** is formed. Between the follicular cells and the oocyte, there forms a membrane called the **zona pellucida** which is made up of an amorphous material and is semipermeable. It is secreted by both follicular cells and oocyte. It is rich in carbohydrate and neutral protein. Sialic acid present in it helps for elasticity of zona Periodic acid-Schiff (PAS +ve).

As the follicular cells are multiplying, small spaces appear between the follicular cells which enlarge gradually and coalesce to form a single cavity called the **antrum folliculi** filled with **follicular fluid or liquor folliculi**. Due to the formation of the cavity, the secondary oocyte, 1st polar body along with the zona pellucida are pushed to the periphery and are attached to the inner aspect of the follicular wall. This is called the mature ovarian follicle or the **Graaffian follicle (Fig. 1.3)**. The secondary oocyte which is in the 1st meiotic division undergoes second meiotic division just before fertilization. If fertilization does not take place, the follicle is extruded along with the menstrual flow before the completion of second meiotic division. The secondary oocyte in 1st meiotic

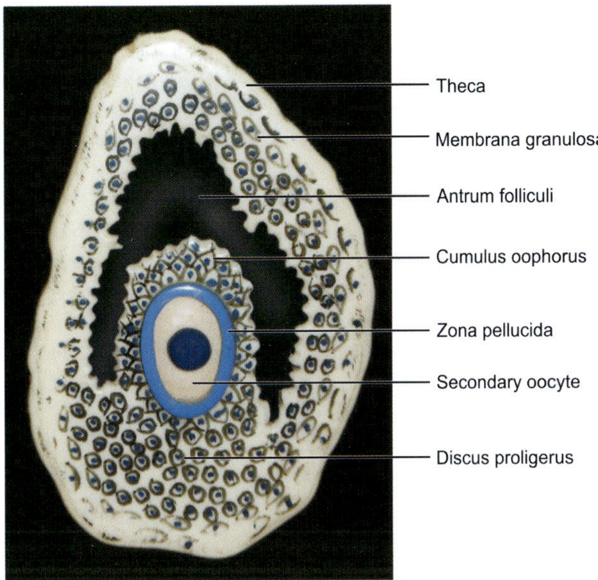

Fig. 1.3: Graafian follicle

division enters into a slow or prolonged 2nd meiotic division and is completed only at the time of fertilization.

The Graaffian follicle consists of secondary oocyte in 1st meiotic division, 1st polar body, both surrounded by zona pellucida. These are attached to the inner wall of the follicle by a group of follicular cells called **discus proligerus**. A layer of cells surround the zona pellucida and are called **cumulus oophorus**. The remaining cells surrounding the antrum folliculi are called the **membrana granulosa**. Around the follicle the stroma condenses and forms **theca**. Theca has an outer fibrous layer **(theca externa)** and an inner cellular layer **(theca interna)**. The theca interna cells and membrana granulosa cells secrete estrogen hormone. The size of the Graafian follicle is about 10 mm diameter and size of the oocyte is about 110 to 140 micrometer in diameter.

Under the influence of luteinizing hormone, the Graafian follicle ruptures and secondary oocyte, 1st polar body, zona pellucida, follicular fluid and the cumulus oophorus (now called corona radiata) cells are released into the peritoneal cavity.

The rupture of the follicle and release of ovum is called **ovulation**.

UTERINE CYCLE (FIG. 1.4)

Throughout the reproductive life of the female a series of cyclical changes occur in the uterine endometrium, ovary and the mucosa of the vagina. The cycle varies from 25 days to 35 days. These cyclical changes are divided into three phases. They are, menstrual, proliferative and secretary phases.

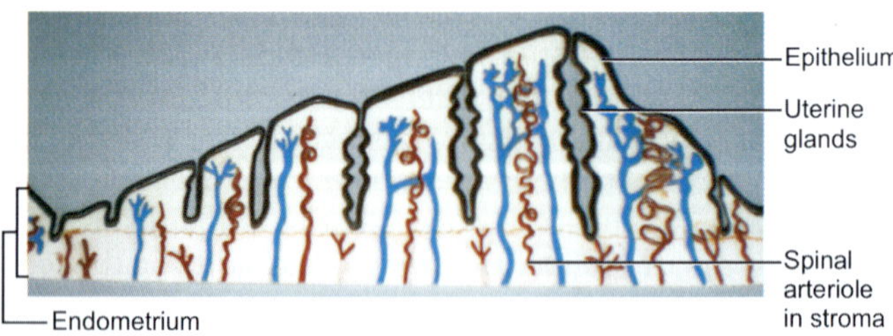

Fig. 1.4: Uterine cycle

Menstrual Phase

Just before menstruation, three strata can be recognized in the endometrium. They are from deep to superficial—stratum compactum, stratum spongiosum and stratum basale.

In stratum compactum, there are necks of uterine glands, compactly arranged stromal cells and connective tissue—all lined on the deeper surface, by simple columnar cells.

In stratum spongiosum, tortuous, dilated uterine glands, stroma in the interglandular space and tortuous capillaries and arterioles. The stromal cells show edematous changes.

In stratum basale, basal parts of glands, arterioles and stroma This layer is thin and lies next to myometrium.

Stratum compactum and spongiosum are collectively called stratum functionale. As the corpus luteum regresses in size, stratum functionale undergoes degenerative changes; endometrium shrinks and gets detached from stratum basale. This stage lasts for 3 to 5 days.

Proliferative Phase

The epithelial lining of the uterine glands in stratum basale grow and replace the detached epithelium. Glands, connective tissue stroma and blood vessels proliferate under the influence of Follicle Stimulating Hormone, progesterone and estrogen. This stage lasts for 10 to 12 days in 28 days cycle.

Secretory Phase

The uterine glands dilate and becomes secretary, secretions are rich in nutrients. The stroma proliferate, blood vessels become tortuous and the cells in the stratum spongiosum become edematous the thickness of the endometrium becomes three times more than that in proliferative phase. This stage is under the influence of progesterone hormone. This stage lasts for 10 to 12 days in 28 days cycle.

In uterine cycle, the proliferative phase shows variation. It is prolonged in longer cycles and reduced in shorter cycles, but the secretory phase remains unchanged and is for 10 days.

CORPUS LUTEUM (CORPORA LUTEA) (FIGS 1.5A AND B)

The term is derived from Latin which means 'yellow body'. It is an endocrine part of ovary which has got short life.

The function of it is to produce hormone called 'progesterone' that is required to maintain the thickness of the endometrium at the time of implantation and growth of embryo.

It is formed when the Graafian follicle ruptures and releases the secondary oocyte (ovulation) on the 14th day of 28 days' menstrual cycle. The rupture of ovarian follicle (Graafian follicle) and formation of corpus luteum are under the influence of luteinizing hormone.

Following the release of the secondary oocyte along with the cumulus oophorus, zona pellucida and follicular fluid, the follicle forms 'corpus hemorrhagicum' due to the rupture of adjacent capillaries and accumulation of blood into the lumen of the ruptured follicle. Then the blood coagulates which is gradually removed by phagocytosis and sprouting of ruptured blood capillaries. These newly growing blood vessels nourish the cells

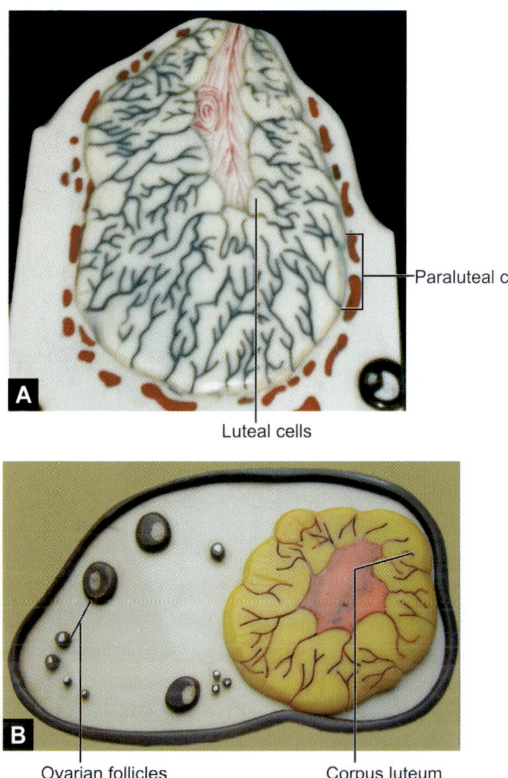

Figs 1.5A and B: Corpus luteum

of membrana granulosa. During this process, the cells enlarge, accumulate yellowish pigment and increase in size. Due to the yellowish appearance, it is called 'corpus luteum' (yellow body).

It is usually large in size and the diameter varies from 2 mm to 5 mm. The yellowish cells are called luteal cells and secrete progesterone. The cells of theca interna of mature ovarian follicle are now called paraluteal cells. These cells do not show much change and continue to secrete estrogen.

Function

The corpus luteum is essential in establishing and maintaining pregnancy by maintaining the growth and thickness of endometrium.

Fate

When the ovum is not fertilized: If the ovum is not fertilized the corpus luteum stops secreting progesterone and decays after approximately 14 days in humans. It is called corpus luteum of menstruation. It later degenerates into corpus albicans, as it appears whitish in color. It is a shrunken mass of fibrous tissue appears as a puckered scar that gives rise to irregular surface to the ovary.

When the ovum is fertilized: If the ovum is fertilized and implantation takes place, the trophoblastic cells secrete human chorionic gonadotropin (hCG), which stimulates the corpus luteum to secrete more of progesterone. Thus, it helps in maintaining the thickness of endometrium. It is therefore called 'corpus luteum of pregnancy'. It normally functions up to third or fourth month of pregnancy and gradually regresses when the placenta takes over the function of secreting progesterone. Sometimes, it remains up to the end of gestation period (ninth month of pregnancy) and regresses to form corpus albicans.

Prostaglandins cause degeneration of corpus luteum of pregnancy and cause abortion of the fetus.

CAPACITATION OR ACTIVATION OF SPERM

Capacitation or activation of sperm is a calcium-dependent event, which involves extensive changes in the sperm.

- The surface membrane covering the head of the sperm fuses at many points with the underlying acrosomal membrane creating a vesiculated appearance and exposing the enzymatic contents of the acrosomal vesicle and also the inner acrosomal membrane to the exterior.
- The tail beat changes from regular wave-like flagellar beats to "whip lashing beats" that push the sperm forward in vigorous lurches.
- The surface membrane of the middle and posterior half of the sperm head fuses to the surface membrane of the ovum.

Destabilization of the surface membranes may be critical.

To summarize:

- Hypermotility of sperms.
- Proteolytic enzymes in female reproductive tract **strip off the glycolytic coat** on the acrosomal cap of the sperm.
- This decreases membrane stability.
- Increases membrane permeability.
- Increases internal calcium levels.
- Increases cyclic adenosine monophosphate (cAMP).

- Increases phosphorylation of proteins.
- Increases tyrosine kinase.

Sperms undergo capacitation in the female reproductive tract.

FERTILIZATION

Following ovulation, the ovum with its cumulus oophorus cells are picked up by the fimbriae of the Fallopian tube.

The ovum has formed the first polar body. It remains in the ampulla portion of the tube and is viable for about 18 to 24 hours.

If fertilization does not occur, the ovum disintegrates and is destroyed by the tube or extruded out along the menstrual flow.

Sperm will remain viable in the female reproductive tract for about 48 hours, although this can be quite variable.

Sperms present in the ampulla meet the cumulus oophorus mass and penetrate by chemical and mechanical means to reach the zona pellucida.

One sperm penetrates the zona pellucida, the second polar body is formed, and the nuclear material of the sperm enters the cytoplasm of oocyte. In the nucleus of secondary oocyte after the completion of 2nd stage of meiosis, chemical changes occur leading to the formation of 'female pronucleus' and the sperm head changes to 'male pronucleus'. Both pronuclei fuse to form the zygote.

The diploid chromosome number is re-established by this fusion and mitotic cell division can now occur.

Sperms stay behind in the oviduct for at least 8 to 24 hours.

Purposes of Fertilization

- Completion of second stage of meiosis in secondary oocyte
- Restoration of chromosomal compliment of the species

- Gives rise to a new individual of its kind
- Determination of the sex of the individual
- At fertilization, a calcium wave is initiated, which also prevents entry of other sperms
- Initiation of cleavage (rapid, successive, mitotic division of zygote).

In Vitro Fertilization

In vitro fertilization is a process by which the ovum is fertilized by sperm outside of the body.

In vitro fertilization or **IVF** involves uniting sperms and ova to create embryos in the laboratory ("*in vitro*"). Once embryos are created, they are placed in the uterus. Thus, the embryos are in uterine cavity growing into normal fetuses.

Procedure

The sperm are separated from the semen by laboratory procedure.

The active sperms are combined in a laboratory dish with ova.

About 18 hours after this fertilization procedure, it is possible to grow as embryo.

Intracytoplasmic Sperm Injection (ICSI)

It is another advanced *in vitro* fertilization where the sperms are injected directly into the cytoplasm of the ovum outside the female reproductive system and then transferred into the uterus after the fertilization.

BLASTOCYST (FIG. 1.6)

After fertilization, the cells divide by mitosis. The cells of the fertilized ovum are called the **blastomeres** and the process of rapid division of cells by mitosis is called the **cleavage**. The cells are all surrounded by zona pellucida. The multiplying cells have the appearance of mulberry and the mass of cells is called the **morula**. As the cell division continues, the morula is propelled slowly into the uterine cavity by peristaltic waves and ciliary beats of columnar cells, of the uterine tube. The secretion of the glandular cells of mucosa provides the fluid vehicle for the transportation and nutrition to the morula.

Between the 5th and 8th day after the fertilization, the morula reaches the uterine cavity (12–16 cell stage).The fluid from the

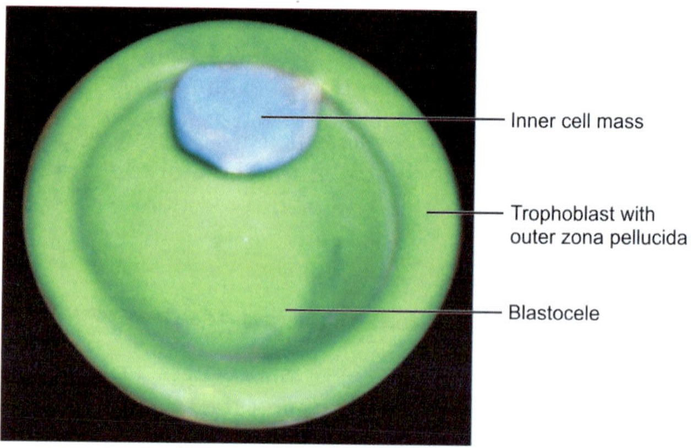

Fig. 1.6: Blastocyst

uterine glands of the endometrium and the secretary cells of the uterine tube diffuse through the zona pellucida to accumulate between the blastomeres in small spaces. The cells in the morula simultaneously differentiate into a central mass and peripheral layer. The spaces filled with the uterine fluid enlarge and coalesce to form a single, large cavity. The **central mass of cells**, also called the **inner cell mass or embryonic mass**, is pushed to the periphery and is attached to the inner surface of the peripheral layer at one end. The peripheral layer is called the **trophoblast**.

The **embryonic mass** of cells gives rise to **embryo proper** and **trophoblast** gives rise to **coverings of the embryo**.

The embryonic mass is separated from the trophoblast by a cavity except at one end and that end is called the embryonic pole. This cystic mass of cells after fertilization is called the **blastocyst**. The cavity of the blastocyst is called the **blastocele** which contains the uterine fluid for the nourishment of the cells of the blastocyst (blastomeres). This stage is called the **blastula**.

The entry of uterine fluid into the blastocele leads to increase in the size of the blastocyst which in turn leads to thinning and disappearance of the zona pellucida and the blastocyst adheres to the uterine mucosa.

TROPHOBLAST

The blastomeres in the fertilized ovum keep dividing by mitosis and undergo differentiation into an inner cell mass and an outer layer. The inner cell mass is the embryonic mass which gives rise to the embryo proper and the outer layer is called the trophoblast which gives rise to the coverings of the embryo. This differentiation takes place at the morula stage and the morula reaches the uterine cavity around 4th or 5th day after the fertilization.

The trophoblastic cells form protective and nourishing membranes of the embryonic mass which gives rise to embryo. The trophoblast cells are smaller than the inner cell mass and rapidly divide, lose their cellular configuration and spread rapidly.

The trophoblast dissolves the uterine mucosa and burrows in the stratum compactum of uterine mucosa. This fast growing trophoblast forms continuous sheet and appears multinucleated because the cells are multiplying before the appearance of the cell membrane. It is called the **syncytiotrophoblast**. The deeper layer of trophoblast cells get back their cell membranes gradually and are called the **cytotrophoblast**.

The trophoblast helps for the implantation of the blastocyst. Due to the irregular growth of the trophoblast, the blood vessels are invaded and blood accumulates in small cavities in the trophoblast which are now called the lacunae. Due to the continuous growth of the trophoblast, there will be formation of finger like processes called **chorionic villi** and the lacunae between them are called the **intervillous spaces**. They contain the **maternal blood**.

The cytotrophoblast grows and invades through the syncytio-trophoblast and is exposed to the stratum basalis of the uterine mucosa. It grows horizontally and completely separates the syncytiotrophoblast from the stratum basalis of uterine mucosa. This cytotrophoblastic covering is called **cytotrophoblastic shell**.

In the later part of pregnancy, the cytotrophoblast disappears and makes the placental membrane thin to facilitate permeability to substances having smaller molecular weight.

IMPLANTATION (FIG. 1.7)

The blastocyst increases in size due to diffusion of uterine fluid and the **zona pellucida thins out and disappears**. The trophoblast now comes in contact with the epithelium of the endometrium which has reached the thickness of 5 mm. The trophoblast cells secrete a sticky substance which also causes lysis of the tissues whichever it comes in contact. By this process of **adhesion, erosion and invasion**, the blastocyst gets inserted into the uterine endometrium. This process of burrowing is called **implantation (Fig. 1.7)**.

Erosion of epithelium, stroma, lining epithelium of uterine glands and the endothelium of blood vessels of the endometrium

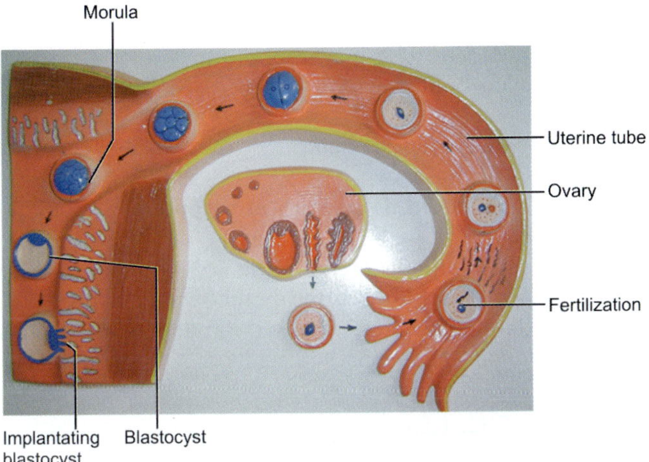

Fig. 1.7: Implantation

takes place and ultimately the trophoblast, which later becomes chorion, comes in contact with the blood of the endometrium. This type of contact is called hemochorial.

Normal Site of Implantation

The normal site of implantation is in the endometrium of the body of the uterus, most frequently in the upper part of the posterior wall near the mid line. But implantation anywhere in the upper half (upper segment) of the uterus is considered normal.

Abnormal Sites of Implantation

- Lower part of the uterine mucosa
- Ampulla of uterine tube
- Isthmus of uterine tube
- Infundibulum of uterine tube
- Ovary
- Peritoneum of broad ligament
- Mesentery of intestine.

The last two could be primary or secondary abdominal implantation.

Sometimes the implantation in the uterine tube or ovary continues up to 6th or 10th week and the tube or the ovary ruptures slowly and developing fetus grows in the abdominal cavity. This is called secondary abdominal implantation.

DECIDUA (FIG. 1.8)

The word 'decidua' is derived from deciduous, that means to '**shed of**'.

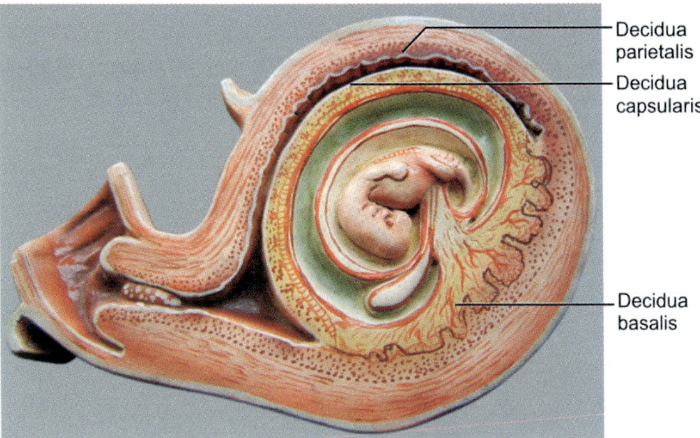

Fig. 1.10: Fate of decidua

Due to the growth of the embryo and the expansion of the cavity of amnion, the decidua capsularis is thinned out, distended and the space between it and decidua parietalis is gradually obliterated. By the third month of gestation, the decidua capsularis and the decidua parietalis are in contact with each other. Gradually in 5th month, the both layers are thinned out and disappear **(Fig. 1.10)**.

The glands in the stratum compactum are obliterated and their epithelium is lost. In the stratum spongiosum, the glands are compressed, appear as oblique slit like fissures and their epithelium degenerates. The epithelium of the glands is retained; they are cuboidal in nature in the basal zone.

TWIN PREGNANCY (FIG. 1.11)

Two fetuses developing in the same uterine cavity simultaneously from same pregnancy is called twin pregnancy.

Multiple Pregnancies

In some cases, the zygote may proceed to an external division and split up into two separate cells. These two new cells then continue

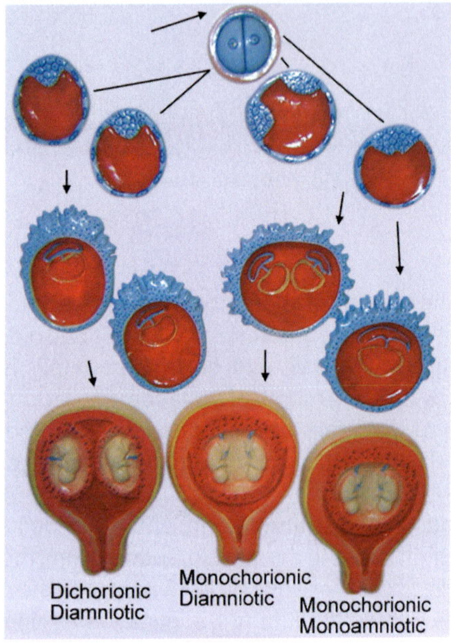

Fig. 1.11: Twinning

their development independently as they undergo their own segmentation. Eventually, they both arrive in the uterus where they also implant separately. This results in multiple pregnancies.

Since both came from the same egg and the same sperm, the twins will be identical, i.e. they will be of the same sex and also look alike in every other respect.

Identical twinning can occur at any time between the second and 14th day after fertilization. Thus, even during implantation the blastocyst may still divide and grow into two separate embryos.

Uniovular or Monozygotic Twins or Identical Twins

During the fertilization process, one ovum is fertilized by one spermatozoon, but the ovum divides into two or more embryos afterwards. They have exactly the same number of chromosomes, have a perfect resemblance and are of the same sex.

Modes of Monozygotic Twinning

There are three different ways that identical twins can develop and be carried:

1. There is only one placenta that feeds the fetuses but there can be two amniotic sacs. When there is 1 placenta and 2 amniotic sacs then the pregnancy is referred to as having a "monochorial" placenta and is "bi-amniotic".
2. There is only one amniotic sac. In the case of 1 placenta and 1 amniotic sac then it is referred to as having a "monochorial" placenta and is "mono-amniotic".

3. There are two placentas and each embryo has its own amniotic pocket. When there are two placentas Monozygotic twins (1/250) and two amniotic pockets—"bi-chorial" pregnancy that is "bi-amniotic".

Binovular or Dizygotic Twins or Nonidentical Twins or Fraternal Twins

When twins are nonidentical, then two separate ova have been fertilized by 2 separate spermatozoa.

They each have a

1. Different chromosome make up,
2. They do not look alike, and
3. They can be either of the same or different sex.

Each embryo is in its own amniotic sac and has its own placenta. This type of pregnancy is referred to as "bi-chorial" and "bi-amniotic".

Zygosity can be determined by DNA fingerprinting. Prenatally, such testing would require an invasive procedure:

1. To sample amniotic fluid (amniocentesis),
2. Placental tissue (chorionic villus sampling) or
3. Fetal blood sampling (cordocentesis).

Determination of chorionicity can be performed by ultrasonography at 6 to 9 weeks of gestation.

In dichorionic twins, there is a thick septum between the chorionic sacs. After 9 weeks, this septum becomes progressively thinner to form the chorionic component of the inter-twin membrane, but it remains thick and easy to identify at the base of the membrane as a triangular tissue projection, or lambda sign.

Different-sex twins are dizygotic and therefore dichorionic, but in about two-thirds of twin pregnancies the fetuses are of the same sex and these may be either monozygotic or dizygotic.

In dichorionic twins, the inter-twin membrane is composed of a central layer of chorionic tissue sandwiched between two layers of amnion, whereas in monochorionic twins there is no chorionic layer between the twins.

There are also some very rare cases of multiple ovulations, i.e. the release of several eggs at the same time. If these eggs are then fertilized, they may produce multiple pregnancies. Twins born as a result of such a pregnancy are known as fraternal twins.

If three of these children are born, they are called triplets, four are quadruplets, five are quintuplets, six—sextuplets, seven—septuplets and so on.

Since each of them comes from a different egg fertilized by a different sperm cell, they may be of different sex and will not resemble each other anymore than other brothers and sisters.

Multiple gestations have higher risks of fetal morbidity and mortality.

Multiple births are common due to the stimulation of ovulation that occurs when exogenous gonadotropins are administered to women with ovulatory failure and those being treated for infertility by drugs, *in vitro* fertilization and embryo transfer.

Chapter 2

Fetal Membranes

Fetal membranes include the following:

- Amnion
- Yolk sac
- Extraembryonic celom
- Allantois
- Connecting stalk/Umbilical cord
- Chorion
- Placenta.

AMNION

Amnion is a thin, tough, fluid filled, membranous sac that surrounds the embryo and the fetus. It is attached to the margins of the embryonic disc.

The amniotic sac contains the **amniotic fluid**, which help keep the baby warm and keeps the baby's growing body parts from fusing together.

To **begin with**, the amniotic fluid is a **transudate** of the maternal plasma and becomes more like the fetal fluids only in the presence of the fetus. When the fetus is removed, the fluid remains similar to maternal plasma.

The origin of amnion is controversial:

Whether it develops from **embryoblast** as **amniogenic cells** and these cells separate out from epiblast and surround the space called amnion or the **trophoblastic cells differentiate into amniogenic** cells or due **delamination of embryoblast and trophoblast from each other** to enclose a slit which becomes amnion.

As amnion enlarges, it gradually obliterates the chorionic cavity (extraembryonic celom) and forms epithelial covering of umbilical cord.

Amniotic fluid plays major role in fetal growth and development.

The fluid is initially secreted by amniotic cells fluid is derived from the maternal tissue and by diffusion across amnio chorionic membrane from decidua parietalis. This is followed by diffusion of fluid through chorionic plate from blood in the intervillous space of placenta.

Before keratinization of the skin, diffusion of water and solutes from fetus to amniotic cavity is through skin, thus the fluid is similar to tissue fluid. The other sources of amniotic fluid are:

- Respiratory tract
- Excreted urine by the fetus (about 300–400 mL).
  By late pregnancy half a liter of urine is added daily.
  The volume of amniotic fluid—30 mL at 10 weeks, 350 ml at 20 weeks. And about a liter by full term.

Composition of amniotic fluid:

- 99% water
- 1% organic and inorganic substances.

Half of the organic constituents are proteins, the remaining half consist of carbohydrate, fat, enzyme, hormones, pigments and desquamated epithelial cells.

Applied Anatomy

- Amniocentesis
- Oligohydramnios
- Polyhydramnios.

YOLK SAC (FIGS 2.1 AND 2.2)

Formation and Boundaries

It is the cavity of the **blastula (blastocele)**, after the formation of the first germ layer called the **hypoblast**. It begins as primordial or **primary yolk sac (Fig. 2.1)**.

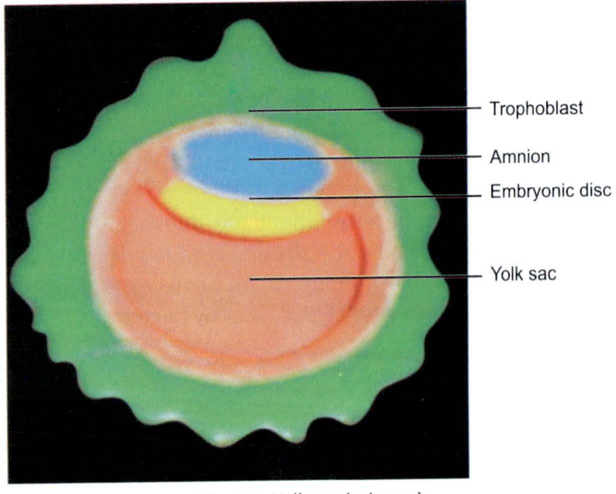

Fig. 2.1: Yolk sac (primary)

The primary yolk sac becomes smaller when the extra-embryonic mesoderm is formed and is called **secondary yolk sac**.

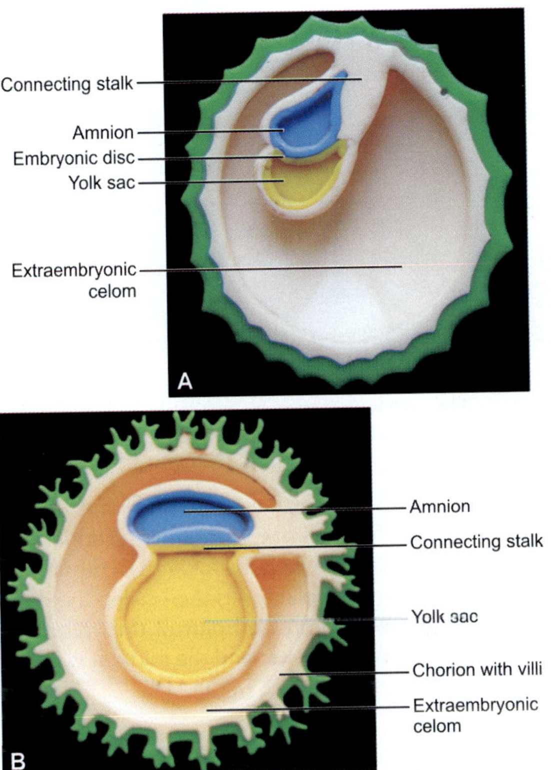

Figs 2.2A and B

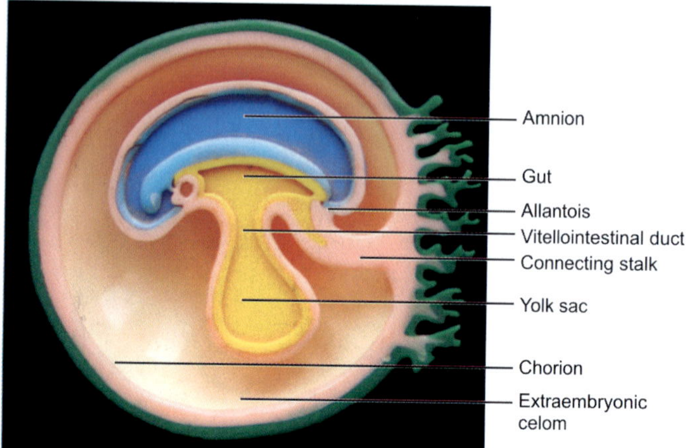

Fig. 2.2C

Figs 2.2A to C: Extraembryonic celom, yolk sac, amniotic cavity and connecting stalk

After the formation of the extraembryonic celom, the secondary yolk sac is further reduced in size. Part of the secondary yolk sac is pinched into the mid gut when the foldings of the embryo and formation of gut tube take place. It is then called the **definitive yolk sac**. The narrow, compressed part of the yolk sac which connects the distal small yolk sac and the mid gut is termed the **vitellointestinal duct (Figs 2.2A to C)**.

Functions

- Nutrition
- Primordial germ cells formation

- Production of blood in the wall of the yolk sac
- Takes part in the formation of primitive gut tube.

Fate of Yolk Sac

- By the 10th week, it is very small and lies between amniotic and chorionic sacs.
- It atrophies as the pregnancy advances.
- In abnormal conditions, the yolk stalk may persist and give rise to **Meckel's diverticulum.**

EXTRAEMBRYONIC CELOM (FIGS 2.2A TO C)

It is the space in the extraembryonic mesoderm, which is also called **chorionic sac.**

Number of small, isolated spaces appear in the mesoderm, increase in size and all of them rapidly fuse to form a large, isolated cavity—the **extraembryonic celom** (EEC).

This fluid filled extraembryonic celom gradually surrounds the embryonic disk, yolk sac and amnion except where they are attached to the chorion by connecting stalk. Due to the increasing size of the extraembryonic celom, the primary yolk sac decreases in size and a smaller secondary yolk sac is formed **(Figs 2.2A to C)**. The EEC splits the extraembryonic mesoderm into an **outer** layer called **extraembryonic somatopleuric mesoderm** and an **inner** layer called the **extraembryonic splanchnopleuric mesoderm.**

Extraembryonic splanchnopleuric mesoderm surrounds the yolk sac.

The extraembryonic somatopleuric mesoderm lines the trophoblast and amnion. Along with trophoblast it forms **'Chorion'**.

Thus, the chorion forms the wall of the chorionic sac within which embryo, amnion and yolk sac are suspended by connecting stalk. Therefore, extraembryonic celom is called chorionic cavity.

Transvaginal ultrasound (endovaginal sonography) is used to measure chorionic sac diameter. This measurement is valuable in evaluating early embryonic development and pregnancy.

Gradually amnion grows in size and expands into extra-embryonic celom.

ALLANTOIS (FIGS 2.3A TO C)

Formation

On 16th day, a sausage shaped diverticulum arises from the **caudal wall of yolk sac**, which extends into the connecting stalk. This is called **allantois**.

It remains very small in human embryos.

During the 2nd month, the extraembryonic part of the allantois degenerates and a portion extends into connecting stalk **(Figs 2.3A to C)**.

The allantois gets incorporated into ventral wall of hind gut after the foldings of the embryo.

The postallantoic part of the hind gut forms cloaca.

Functions

- Blood formation between 3rd and 5th week of IUL.
- Development of umbilical vessels.
- In reptiles, birds and mammals it has respiratory function and acts as reservoir for urine.

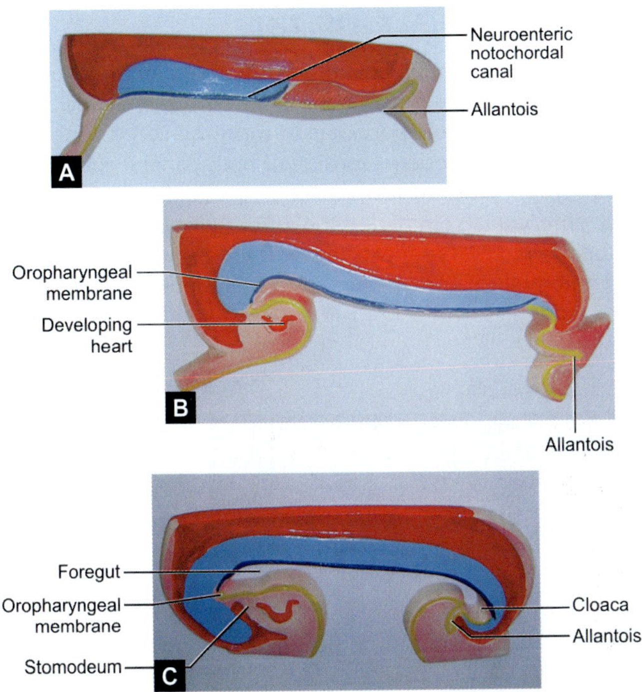

Figs 2.3A to C: Allantois

Fate

It extends between urinary bladder and umbilicus. As the bladder enlarges, the allantois involutes to form urachus. Later on, the urachus degenerates to form median umbilical ligament.

CONNECTING STALK (FIG. 2.4)

On 14th day, the embryonic disc with its amnion and yolk sac becomes suspended in the chorionic cavity by a thick layer of mesoderm, which elongates to form the connecting stalk **(Fig. 2.4)**. The mesoderm consists of both parietal and visceral extra-embryonic mesoderm. It is gradually lined by the extraembryonic celomic epithelium. It elongates and forms umbilical cord that connects the fetus to the mother.

The umbilical cord begins to form in about 5 weeks after conception.

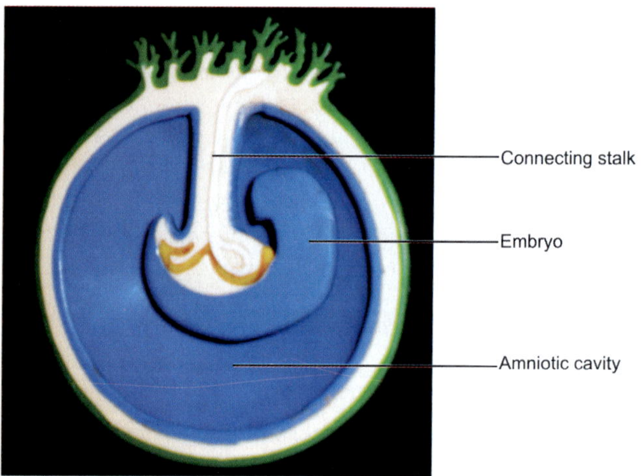

Fig. 2.4: Amniotic cavity and connecting stalk

UMBILICAL CORD

The connecting stalk becomes progressively longer until about 28 weeks of pregnancy, reaching an average length of 22 inches. As it gets longer, the cord generally twists around itself and becomes coiled. Gradually the allantois gets incorporated. The umbilical vessels, vitelline vessels are incorporated. The mesoderm will have contribution from both intraembryonic and extraembryonic mesoderm.

It is the cord that connects the fetus to the mother. It is a twisted cord, which carries blood to and from the placenta.

Length: 50 cm (Ranges from 20 cm to 120 cm).

Diameter: 1 to 2 cm.

The umbilical cord is a narrow, tube-like structure that connects the developing fetus to the placenta.

The umbilical cord delivers nutrients and oxygen for the normal growth and development. It also removes waste products.

It consists of:

An outer covering of amniotic membrane.

An interior mass of mesoderm derived from:

- Somatopleuric extraembryonic mesoderm covering amniotic membrane.
- Splanchnopleuric extraembryonic mesoderm of yolk sac.
- Splanchnopleuric extraembryonic mesoderm of allantoic duct and umbilical vessels.
- Yolk duct and allantoic duct.
- Vitelline and umbilical vessels.
- Remains of extraembryonic celom.

The amniotic membrane is made up of epithelial cells which are simple squamous or cuboidal cells.

The mesodermal core, from different sources, fuses and gradually transformed into viscid, mucoid connective tissue which is called 'Wharton's jelly'.

It consists of:

- Widely spaced fibroblasts (with branching processes) separated by extensive intercellular space which is filled with
- Copious matrix consisting of delicate, three dimensional meshwork of fine collagen fibers surrounded by
- Dilute ground substance (with a variety of hydrated mucopoly-saccharides).

Yolk and allantoic ducts slowly degenerate and disappear. Only interrupted cords of cells may be seen.

Vitelline vessels atrophy in the region of the umbilical cord.

Umbilical vessels are—2 arteries and 2 veins—right and left.

The right umbilical vein disappears in the early months of gestation.

The vessels are twisted- in right or left handed cylindrical helix **(Fig. 2.5)**.

The number of twists or turns may exceed 300.

They are due to unequal growth of vessels and torsion forces imposed by fetal movements.

The arteries are medium sized, muscular arteries.

The external layers of tunica media of the arteries pursue oblique/spiral course.

Contraction produces shortening of vessels and thickening of tunica media. The pulsations of the umbilical arteries assist venous return to fetus through umbilical vein. This leads to folding/bulging of tunica intima which narrows the lumen and act as valves during pulsation.

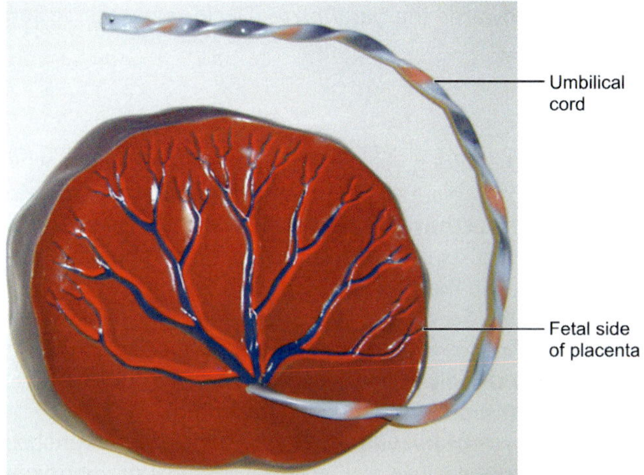

Fig. 2.5: Umbilical cord and placenta

These foldings are called 'Valves of Hoboken'.
Cord exhibits knots. They are:

- **True knots**—due to fetal movements.
- **False knots**—due to sharp variation in contour due to excess accumulation of Wharton's jelly.

CHORION

Formation

The trophoblast and the somatopleuric extraembryonic mesoderm is called the 'chorion'.

It grows towards the basal plate (Decidua basalis) to form placenta.

Parts

It consists of the thick part called the **chorion frondosum**, which is nearer the basal plate and the thin, smooth part called **chorion leavae which is at the region of decidua capsularis.** The chorion frondosum consists of thickly arranged branching, finger like projections called the **chorionic villi**.

Chorionic villi are of three types depending upon the content in the wall:

1. **Primary chorionic villi**—Contain only cytotrophoblast (green) and syncytiotrophoblast (blue) **(Fig. 2.6)**.
2. **Secondary chorionic villi**—Contain outer syncytiotrophoblast, inner cytotrophoblast and a central core of extraembryonic, somatopleuric mesoderm (light pink) **(Fig. 2.7)**.

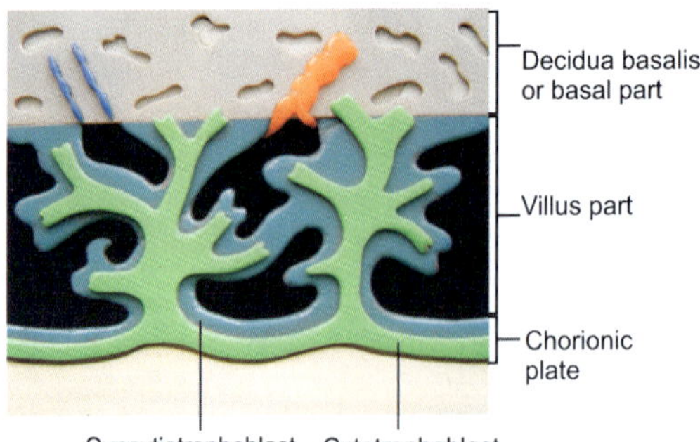

Fig. 2.6: Primary chorionic villi

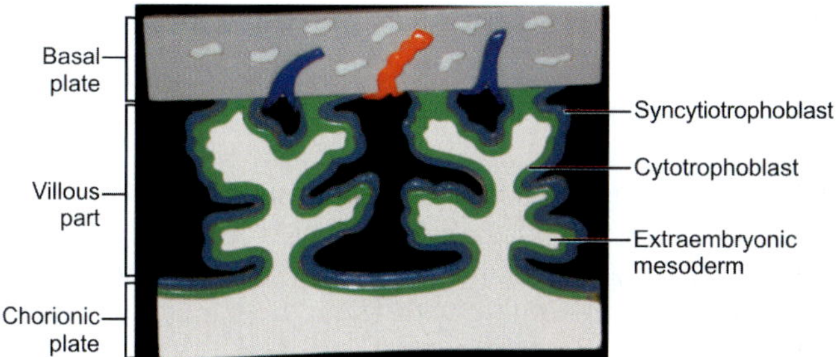

Fig. 2.7: Secondary chorionic villi

3. **Tertiary chorionic villi**—Outer syncytiotrophoblast, deeper cytotrophoblast, inner extraembryonic mesoderm and the fetal capillaries **(Fig. 2.8)**.

Depending upon the attachment of the villi to the basal plate the chorionic villi are classified into:

- Anchoring villi
- Floating villi.

According to the branching pattern:

- Truncus chorii
- Rami chorii
- Ramuli chorii.

Depending upon the thickness and amount of the chorion and villi the chorion consist of:

- Chorion frondosum (thicker part)
- Chorion leavae (thin, smooth part).

Fig. 2.8: Tertiary chorionic villi

PLACENTA (FIGS 2.8 AND 2.9)

The word 'Placenta' is derived from 'plakuos' a Latin word which means 'Flat Cake'.

It is the organ, characteristic of true mammals during pregnancy, joining mother and fetus, provides nutrition, hormones and removal of the waste from the fetus.

Full-term Placenta

Definition

It is a fetomaternal tissue which is discoidal, deciduate, hemochorial, chorioallantoic, initially labyrinthine, later a villous organ.

Fig. 2.9: Placental circulation

- Shape is disc like (discoidal)
- Is shed off (deciduate)
- Fetomaternal connection is chorion and blood (hemochorial)
- It is vascularized by allantoic vessels of body stalk (chorioallantoic).

The trophoblast has interconnecting tunnels (labyrinthine), which become finger like later (villous).

Measurements of Full-term Placenta (See Fig. 2.5)

- Diameter = 185 mm (range is 150–200 mm)
- Weight = 500 g (range 200–800 g)
- Thickness = 23 mm (range = 10–40 mm)
- Surface area = 30,000 sq mm
- Thickest at the center
- Has two surfaces—fetal and maternal.

Fetal surface is smooth, contains blood vessels and is covered by amniotic membrane. Umbilical cord is attached in the middle **(Fig. 2.9)**.

The maternal surface is finely granular and mapped into 15 to 20 irregular, polyhedral lobes separated by fissures or grooves.

The grooves are the sites of placental septa.

The lobes are called 'Cotyledons' or 'Uterine caruncles' (each cotyledon is called as lobe by obstetricians).

Fetal unit: Area supplied by major umbilical vessels along with corresponding maternal tissue called placentome. Fetal cotyledon corresponds to major villus stem and its branches.

Early in pregnancy the chorion bears 800 to 1000 such stems but as gestation advances, due to disappearance and fusion of remaining persisting stems, the number is reduced to 60 which are distributed between 15 and 20 lobes.

Therefore, each lobe contains 2 to 4 major villus stems. The cells of the villi are rich in organelles and inclusion bodies indicating highly active cells. As gestation advances the endothelial cells and syncytiotrophoblast cells become thin and elongated.

Microscopic Picture of Placenta

From outside inwards:

- Basal plate
- Chorionic plate
- Villous part containing chorionic villi and intervillous spaces filled with maternal blood.

Basal Plate

It is made up of, from fetal to maternal side:

- Syncytium—patches of maternal tissue which forms the outer wall of intervillous space

- Rohr's stria of fibrinoid
- Remains of cytotrophoblastic shell
- Nitabuch's stria of fibrinoid
- Remains of decidua—ramnants of uterine glands, decidual cells and blood vessels plexus
- Giant cells of fetal origin—grouped syncytiotrophoblast.

Chorionic Plate

From fetal aspect:

- Amniotic epithelium
- Connective tissue layer with main branches of umbilical vessels—derived from mesoderm
- Diminishing layer of cytotrophoblast
- Inner syncytial wall.

These are the essential structures for the exchanges between the mother and the fetus, without mixing the maternal and the fetal blood.

This layer, which separates both blood is called 'placental membrane' or 'placental barrier'.

The thickness of the placental membrane or placental barrier in the initial stages of gestation is about 0.025 mm.

As the gestational age advances, the membrane becomes thinner due to disappearance of mesoderm and cytotrophoblast and also thinning and elongation of endothelium and syncytiotrophoblast, it becomes 0.002 mm.

Endothelial layer rests on basement membrane.

The mesoderm consists of fibroblasts, Hofbauer cells, plasma cells, delicate collagen fibers.

Hofbauer cells are more in early stage and are phagocytic cells.

Gradually, the mesodermal cells disappear leaving only the fibrous network.

Cytotrophoblast (Langhans Cells)

These cells are pale, light basophilic, initially form continuous layer of cells with distinct cell membrane for each cell.

- Gradually these cells disintegrate and disappear.
- The gaps between cells is filled with fibrinoid.
- In the later part of gestation, these cells disappear.

Syncytiotrophoblast

There is no prominent cell membrane; therefore, there is continuity of cytoplasm between adjacent cells.

Cytoplasm is strongly basophilic.

These cells are rich in organelles and inclusion bodies, indicating highly active cells.

As gestation advances, the cells become thin and elongated.

The syncytial growth are seen inside the villus and outside the villus.

Inside the villus (ingrowths): Seen towards mesodermal core – stromal trophoblastic buds. May approach and appear in fetal vessels and are found in umbilical vein.

Outside the villus (outgrowths): Into intervillous space are called syncytial knots or clumps which represent a sequestration phenomenon.

Fusion of knots-bridges, form struts for support of the villus.

There are syncytial sprouts which continue to form throughout pregnancy. They are initial parts of new branches of villi.

They project into intervillous space and often get detached forming maternal syncytial emboli which pass to lungs.

There are about 100,000 sprouts into maternal circulation, but disappear by lysis.

Occasionally form a locus for neoplastic growth.

Development of Third Week of Life

Chapter 3

The following structures are formed in third week:

- Primitive streak
- Notochord
- Neural tube
- Neural crest
- Somites
- Intraembryonic celom
- Blood vessels and blood of the fetus
- Completion of chorionic villi development
- Prochordal plate (buccopharyngeal membrane) and cloacal membrane.

PRIMITIVE STREAK

The **primitive streak** is a **linear band of thickened epiblast** that first appears at the **caudal** end of the embryo and **grows cranially**. At the cranial end, its cells proliferate to form the **primitive knot** (**primitive node)**. With the appearance of the primitive streak, it is possible to distinguish cranial (primitive knot) and caudal (primitive streak) parts of the embryo.

Primitive knot and the streak are formed in the following way:

- **Cell proliferation**—Causes heaping up of the cells caudal to the prochordal plate and epiblast is the source of a new layer of cells.

- **Cell migration** by amoeboid movement—The cells insinuate themselves between the epiblast and hypoblast.
- **Cell determination**—The cells arising from the primitive streak are determined to give rise to different rudiments.

The cells of the primitive streak which insinuate between the epiblast and secondary endoderm extend all over the embryo except the prochordal plate cranially and cloacal membrane caudally. Thus, a third layer is formed which is called the intraembryonic mesoderm (secondary mesoderm) **(Figs 3.1 and 3.2)**. By this process, cells of the epiblast are translocated to new positions in the embryo, producing the three primary germ layers. Formation of the secondary or intraembryonic mesoderm is called the Gastrulation.

Gastrulation is the crucial time in the development of multicellular animals. During gastrulation, several important things are accomplished:

- The three primary germ layers are established.
- The basic body plan is established, including the physical construction of the rudimentary primary body axes.

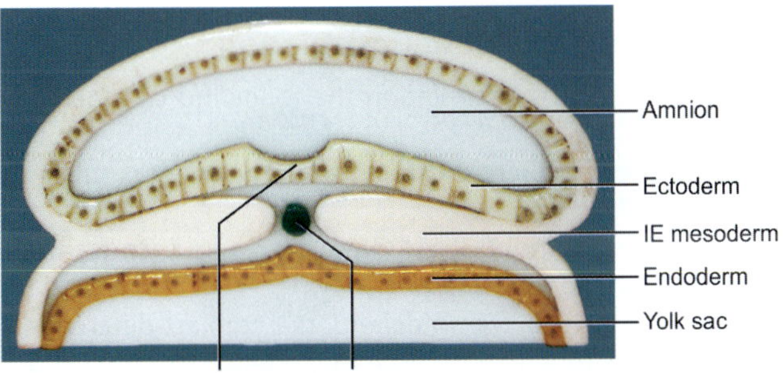

Fig. 3.1: Intraembryonic mesoderm

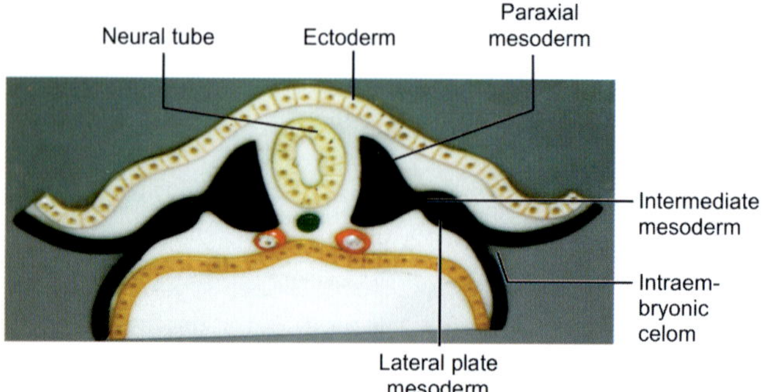

Fig. 3.2: Fate of intraembryonic mesoderm

As a result of the movements of cells in gastrulation, the cells are brought into new positions, allowing them to interact with cells that were initially not near them. This paves the way for inductive interactions, which are the hallmark of neurulation and organogenesis.

Changes Consequent to Gastrulation

With the process of gastrulation the following changes have occurred:

- The embryo becomes **a trilaminar embryo** but is still in the form of a flat disc.
- Epiblast and hypoblast are now known as **ectoderm** and **secondary** endoderm respectively.
- With the formation of notochord, the epiblast differentiates into neuroectoderm and surface ectoderm.

- Hypoblast is replaced by epiblast cells that forms secondary endoderm.
- Mesoderm does **not** extend between epiblast and secondary endoderm at the prochordal plate (buccopharyngeal membrane) and cloacal membranes.
- Intraembryonic mesoderm merges with the extraembryonic mesoderm at the periphery of the embryonic disc.
- Gastrulation converts the embryo into a trilaminar disc.
- Gastrulation can be inhibited by drugs that interfere with actin filament formation, which inhibit cytokinesis and cell migration.
- The intraembryonic mesoderm differentiates into paraxial mesoderm, intermediate mesoderm and lateral plate mesoderm.
- Intraembryonic celom is formed in the lateral plate mesoderm which splits the lateral plate mesoderm into somatopleuric and splanchnopleuric intraembryonic mesoderm.
- The splanchnopleuric intraembryonic mesoderm cranial to prochordal plate is called cardiogenic mesoderm in which heat develops.
- The mesoderm at the cranial end of the embryo where the somatopleuric and splanchnopleuric intraembryonic mesoderm are continuous is called septum transversum that contributes to diaphragm and liver.
- Paraxial mesoderm forms axial skeleton and muscles attached to it.

  Intermediate (nephrogenic) mesoderm forms excretory system.

DEVELOPMENT OF NOTOCHORD

It is the forerunner of axial skeleton.

It is the primary inductor in the early embryo—induces overlying ectoderm to form neural plate.

Notochord develops as follows:

A solid cord of cells grow from the cranial end of the primitive knot towards the prochordal plate. This process is called the head process or notochordal process.

This process pierces the endoderm just caudal to the prochordal plate and comes in contact with the yolk sac.

The notochordal process now undergoes canalization from the continuation of a depression, called blastopore, at the center of the primitive knot.

The canal is called notochordal canal (neurenteric canal) which connects the amniotic cavity with the yolk sac.

This connection helps in nourishing the endoderm and its derivatives till the blood supply to the deeper structures is established.

The ventral aspect of the notochordal canal breaks at multiple points and the canal communicates with yolk sac at multiple levels.

Gradually the floor (ventral wall) of the notochordal canal disappears and notochordal plate followed by notochordal groove is formed.

The ends of the notochordal groove separate from the endoderm and join to form the notochord.

The separated ends of the endoderm now join to form a continuous endodermal layer. The connection between the amniotic cavity and the yolk sac will be closed as the notochord is formed.

The notochord degenerates to form nucleus pulposus of intervertebral disc. It is incorporated into the apical ligament of dens, basiocciput and basisphenoid in the adult life.

FORMATION OF NEURAL TUBE

The neuroectoderm is derived from the epiblast and is induced by the underlying notochord during the third week. It extends along the axis of the embryo just dorsal to the notochord. It is called the **neural plate (Fig. 3.3A)**.

Late in the third week, the neural plate begins to fold. It is first converted into a **neural groove (Fig. 3.3B)**. The lateral edges of the groove are raised to form neural folds.

The neural groove deepens and eventually forms a **neural tube (Figs 3.3C and D)**.

Two masses of ectoderm along the edges of the neural plate form the **neural crest (Figs 3.3B to D)**.

Initially the neural crest separates neuroectoderm from skin ectoderm.

As folding of the neural tube occurs, the neural crest cells detach from the ectoderm and neural tube and form clusters that migrate into the mesoderm.

This process of formation of neural tube and neural crest is called the Neurulation.

The neural tube elongates, undergoes folding called the flexures and cranial part dilates at three parts separated by constrictions which are called the primary brain vesicles **(Fig. 3.4A)**. These primary brain vesicles undergo further subdivisions into secondary brain vesicles which give rise to various parts of the brain. Rest of the neural tube caudal to the brain vesicles forms the spinal cord **(Fig. 3.4B)**.

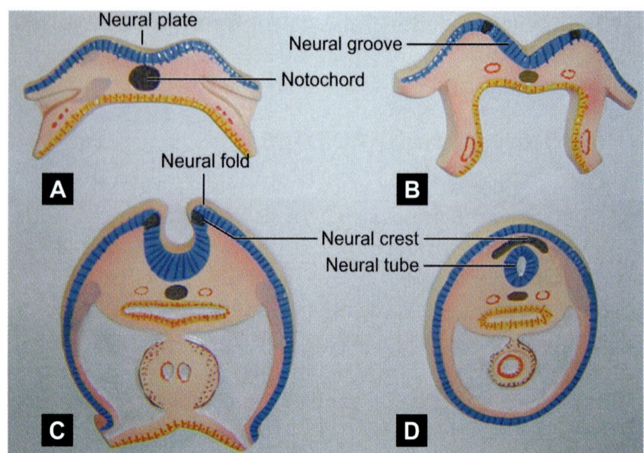

Figs 3.3A to D: Formation of neural plate, neural groove, neural tube and neural crest

The neural tube consists of anterior (cranial) and posterior (caudal) ends called anterior (cranial) neuropore and posterior (caudal) neuropore. The anterior neuropore closes by 24th or 25th day after fertilization and is represented by lamina terminalis. The posterior neuropore closes by 26th or 27th day after fertilization.

The primary inductor for the formation of neural ectoderm is the notochord.

NEURAL CREST

The neural crest is a specialized subpopulation of ectodermally derived cells originating at the border between the neural plate

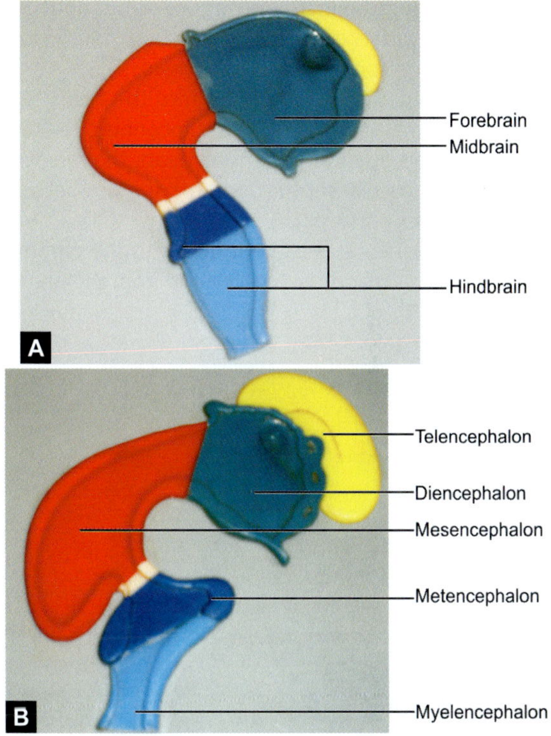

Figs 3.4A and B: Neural tube

and future epidermis in the dorsal aspect of all vertebrate embryos **(Fig. 3.3D)**.

Neural crest cells delaminate and migrate throughout the body and contribute to variety of tissues ranging from neurons and glia to connective tissue components of the head.

The cardiac neural crest migrating into the heart region is responsible for forming the entire musculoconnective tissue wall of the large arteries emerging from the heart, the membranous portion of the ventricular septum, and the septum between the aorta and pulmonary artery.

In addition, the neural crest contributes to the parathyroids, thyroid, and thymus glands which develop from the pharyngeal apparatus **(Fig. 3.5)**.

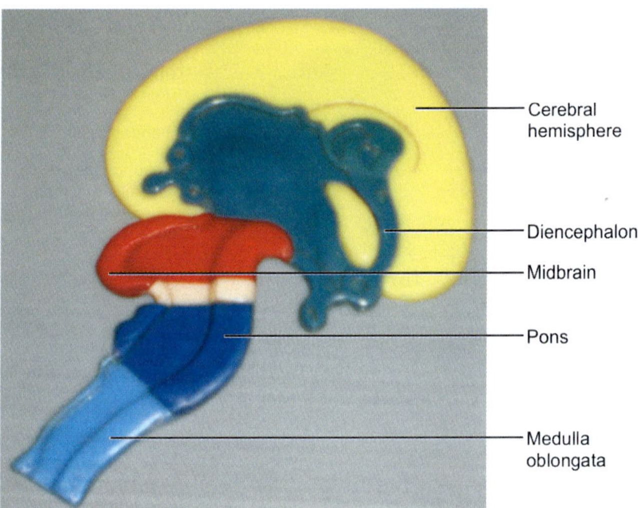

Fig. 3.5: Neural tube derivatives

Neural crest derivatives are:

- Melanocytes
- Ganglia of dorsal root of common spinal nerves
- Schwann cells
- Adrenal medulla
- Sensory ganglia of V, VII, VIII, IX, X cranial nerves
- Autonomic ganglia
- Enamel of tooth
- Para follicular cells of thyroid
- Cells of Pia mater
- Cells of arachnoid mater
- Bipolar cells of sensory epithelium
- Multipolar neurons
- Cartilage and connective tissue of the third, fourth and sixth pharyngeal arches.

SOMITES

As the notochord and neural tube develop, the intraembryonic mesoderm along the lateral side forms longitudinal columns called **paraxial mesoderm.** These columns divide by metameric segmentation into paired cubical bodies called **somites (Fig. 3.6).** The somites develop in pairs; the first pair develops near the cranial end of the notochord around the end of the third week. Additional pairs of somites develop in a caudal direction from days 20 to 30. Totally about 44 pairs of somites appear in the embryo. The first pair it is called **period of somites development** or somatic period and the number of somites is sometimes used as a criterion for determining the age of embryo. The somites give rise to most

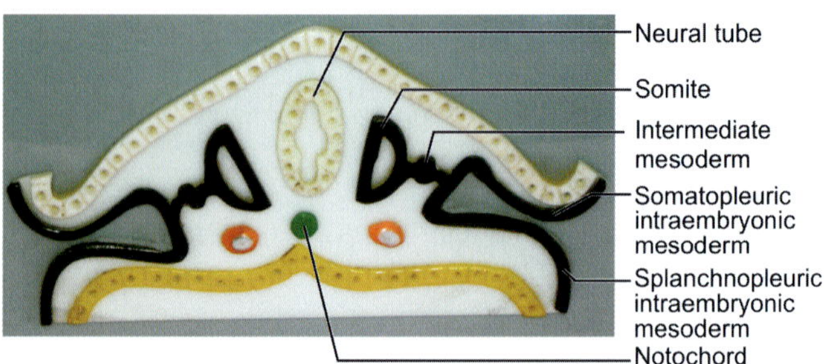

Fig. 3.6: Somites

of the axial skeleton like vertebral column, ribs, sternum, and skull base and associated musculature, as well as to the adjacent dermis.

Somite is divided into two parts:

1. The ventromedial part—**sclerotome**. It contains a "cavity" of loose cells. Cells from the sclerotome migrate medially to surround the notochord and neural tube and form the axial skeleton.
2. The dorsolateral part—**dermomyotome**. Cells from the dermomyotome migrate laterally and, as its name implies, gives rise to (i) skeletal muscle and (ii) the dermis of the skin.

The myotome contains a cavity in the early stages called 'myocele' which obliterates immediately. The concept of myotome in gross anatomy is an embryological concept. Each anatomical myotome is derived from the embryological dermomyotome that is innervated by a segmental nerve. It forms a group of skeletal muscles and the dermis of the corresponding segment of skin.

EMBRYONIC FOLDING

Folding occurs by differential growth of tissues. Neural ectoderm grows faster than the surrounding skin ectoderm and consequently folds to form a neural tube. Similarly, skin ectoderm grows faster than the underlying mesoderm and endoderm, and this differential growth causes folding of the trilaminar disc and gives shape to the embryo.

This period is also called **period of organogenesis**.

It is the critical period of embryo **(Figs 3.7 and 3.8)**.

All major external and internal structures are established during this period.

The main organ systems have begun to develop but function of most of them is minimal except cardiovascular system.

As the tissue and organs form, the shape of the embryo changes and by the end of 8th week, it has a distinct human appearance.

Phases of Embryonic Development

- First phase—Growth.
- Second phase—Morphogenesis.
- Third phase—Differentiation.

Fig. 3.7: Sagittal section of embryonic disc before folding

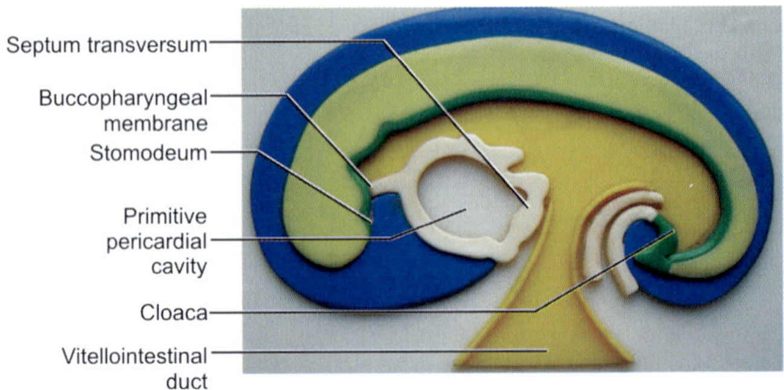

Fig. 3.8: Sagittal section of embryonic disc showing foldings of the embryo

Since the tissues and organs are differentiating during the period, exposure of the embryo to teratogens may cause major congenital anomalies.

Folding occurs mainly at the edges of the embryonic disc around a firm/stiff midline parts formed by neural tube, notochord and somites and forms three main folds **(Fig. 3.8)**:

1. Head fold
2. Tail fold
3. Lateral folds—convert the embryo into a tubular structure.

These are not three separate folds but occur simultaneously and merge into one another.

As a result of the formation of the **head fold**:

- The **foregut** is formed by folding of the endoderm
- The **stomodeum** is an invagination of ectoderm, and has the buccopharyngeal membrane separating it from the foregut. It opens into the amniotic cavity.

- The **pericardial cavity** and **cardiogenic mesoderm** are shifted to the ventral aspect of the embryo and lie **ventral** to the foregut.
- The part of the transverse mesoderm between the pericardial cavity and the yolk sac is the **septum transversum** proper. The liver develops in this **(Fig. 3.8)**.
- The **amniotic cavity** extends ventral to the cranial end of the embryo.
- The **yolk sac** is constricted from the cranial aspect.

As a result of the formation of the **tail fold**:

- The hindgut is formed.
- The connecting stalk is shifted ventrally.
- The allantoic diverticulum is shifted ventrally. It is an invagination of hindgut endoderm into the yolk sac.
- The amniotic cavity extends ventral to the caudal end of the embryo.

  The cloaca is an invagination of ectoderm and has the cloacal membrane separating it from the hindgut.
- The yolk sac is constricted from the caudal end.

Transverse or Lateral Folding of the Embryo

- Converts the endoderm into a primitive gut tube
- The intraembryonic celom surrounds the gut tube
- The communication between the intra- and extraembryonic celoms becomes constricted and eventually obliterated.

Important changes occur in the embryonic cavities as a consequence of folding:

- The **amniotic cavity** surrounds the embryo completely on all aspects and becomes the predominant cavity. It enlarges progressively **(Fig. 3.8)**.

- The **yolk sac** becomes constricted on all sides and part of it is absorbed into the midgut. It becomes a small sac connected to the midgut by a narrow vitelline duct. It becomes progressively smaller.
- The **extraembryonic celom** is gradually obliterated by the expanding amnion and eventually disappears completely.

DERIVATIVES OF GERM LAYERS

Ectoderm

- Central nervous system
- Peripheral nervous system
- Sensory epithelia of the eye, ear, and nose
- Mammary glands
- Enamel of teeth
- Epidermis and epidermal derivatives of the integumentary system including the hair follicles, nails and the glands that communicates with the skin surface
- The lining of the vestibule of the oral cavity
- Parotid salivary glands
- Nasal passageways
- Anal canal below pectinate line
- Portions of the skull, pharyngeal clefts
- Anterior pituitary
- Medulla of adrenal glands
- Bone of pharyngeal (branchial) arch origin
- Ciliaris muscle of eyeball

- Arrectores pilorum muscle of skin
- Melanocytes.

Endoderm

- Epithelial lining of the gastrointestinal tract
- Epithelial lining of respiratory tracts
- Epithelia lining of the tympanic cavity
- Epithelial lining of mastoid antrum
- Epithelial lining of auditory tube
- Epithelial lining of the urinary bladder and most of the urethra
- Liver
- Pancreas
- Parathyroid glands
- Thymus
- Thyroid
- Tonsils.

Mesoderm

- Blood
- Bone
- Cartilage
- Connective tissue
- Cortex of adrenal glands
- Genital ducts (mesonephric and paramesonephric ducts)
- Heart
- Kidneys
- Lymphatic vessels

- Ovaries
- Testes
- Serous membranes of the body cavities (pericardium, pleura, and peritoneum)
- Spleen
- Striated and smooth muscles
- Dermis of the skin.

Developments in Fourth Week and Onwards

BRANCHIAL APPARATUS

Around 4th week of gestation, as the neural crest cells migrate into future head region, the pharyngeal apparatus begins to develop.

This apparatus develops in the future pharynx therefore called 'Pharyngeal Apparatus' **(Fig. 4.1)**.

It appears in the region of the visceral mesoderm around the cranial part of foregut, which is surrounded by endoderm inside and ectoderm outside, therefore it is also called 'Visceral Apparatus'.

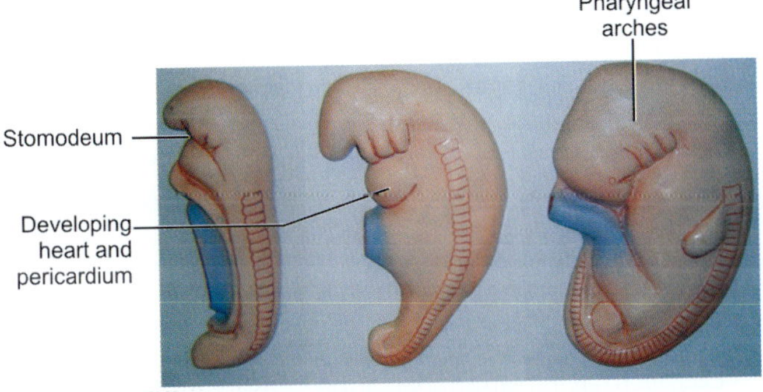

Fig. 4.1: Pharyngeal apparatus

This region of the foregut represents the gills of the fish, therefore also called 'Branchial Apparatus' (branchia, in Greek, means Gills).

Between the stomodeum and developing heart, in the visceral mesoderm of the foregut, horizontal thickenings appear (initially the visceral mesoderm later by neural crest).

These horizontally thickened bars grow forwards/anteriorly and fuse with each other in the anterior midline to form **pharyngeal/ branchial arch**.

Thus, six such arches appear which displace the heart downwards. Later the 5th arch disappears.

Each arch has a mesodermal core covered externally by ectoderm and internally by endoderm.

Internally, the endoderm grows between adjacent arches in the form of diverticulum and is called **pharyngeal/branchial pouch**. There are 4 pouches and 5th pouch is rudimentary.

Externally, the ectoderm, presents shallow depressions between the adjacent arches, which are termed the **pharyngeal/ branchial clefts**. Four clefts are present.

Between the arches, the ectoderm and endoderm are separated by a thin layer of mesoderm which is called branchial plate or pharyngeal membranes.

In the fish, these branchial plates rupture to form gill slits. In the human, they do not rupture and are soon invaded by the mesoderm of the neighboring arches.

PHARYNGEAL ARCHES (FIG. 4.2)

Derivatives of **pharyngeal arches** are listed in **Table 4.1**.

Each arch has its own arterial supply, sensory and motor nerves and skeletal elements.

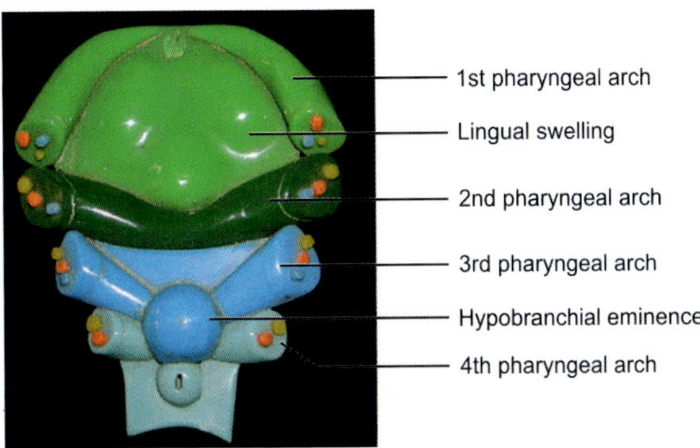

Fig. 4.2: Pharyngeal arches

First arch has two nerves—one along the cranial border (post-trematic) and another along the caudal border (pretrematic nerve). The remaining arches have one nerve along the cranial border (Trema = Cleft).

The arteries of the arches are mentioned in the 'Aortic arches and their derivatives' section (*See* **Figs 4.19 and 20**).

PHARYNGEAL POUCHES

The internal surfaces of pharyngeal arches are lined by the endoderm of foregut.

Endodermal diverticula between pharyngeal arches are called pharyngeal pouches **(Fig. 4.3)**.

Derivatives of Pharyngeal Pouches

I. **Pharyngeal pouch:** Ventral part—Mucosa of tongue and pharynx.

Table 4.1: Derivatives of pharyngeal arches

Arch	*Nerve*	*Derivatives in the adult*	
		Muscle	*Skeletal derivatives*
First (Mandibular)	Post-trematic—Mandibular branch of Vth cranial nerve Pretrematic—Chorda tympani branch of VIIth cranial nerve	Muscles of mastication, Tensor palatine, tensor tympani, mylohyoid, anterior belly of digastric muscles	Malleus, incus, mandible, sphenomandibular ligament, anterior ligament of malleus
Second (Hyoid)	Post-trematic—VIIth cranial nerve	All the muscles of facial expression, auricular muscles, scalp muscles, stapedius, stylohyoid muscle, posterior belly of digastric muscle	Stapes, styloid process of temporal bone, lesser cornu and upper part of the body of hyoid bone. Stylohyoid ligament
Third	Post-trematic—IXth nerve	Stylopharyngeus muscle, constrictor muscles of pharynx (?)	Greater cornu and lower part of body of hyoid bone
Fourth	Post-trematic—Superior laryngeal branch of vagus (Xth cranial) nerve	Muscles of soft palate except tensor palatine, muscles of pharynx except stylopharyngeus, cricothyroid	Laryngeal cartilages
Sixth	Post-trematic Recurrent laryngeal nerve	Intrinsic muscles of larynx except cricothyroid	Laryngeal cartilages

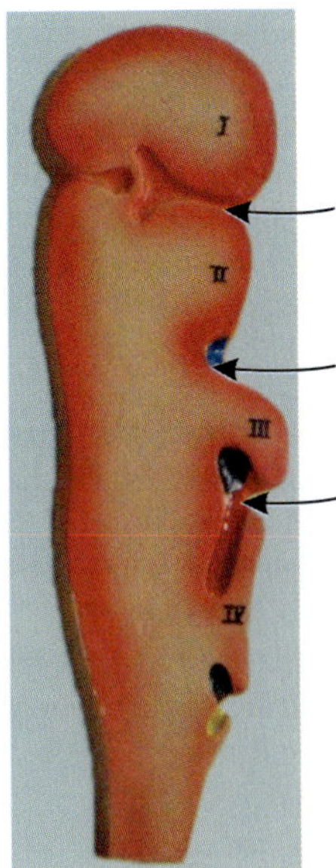

Fig. 4.3: Pharyngeal pouches

Dorsal parts of I and II pouches—tubotympanic recess—forms, mastoid antrum, middle ear and auditory tube.

II. Pharyngeal pouch: Ventral part—mucosa of tonsil.
Dorsal part—contributes to tubotympanic recess—forms middle ear, mastoid antrum and auditory tube.

III. Pharyngeal pouch: Ventral—thymus—epithelioreticular cells.
Dorsal—inferior parathyroid.

IV. **Pharyngeal pouch:** Ventral—thyroid (lateral thyroids). Dorsal—superior parathyroid.
(Rudimentary V pouch gives rise to ultimobranchial body or caudal pharyngeal complex. Neural crest cells migrate into this body and gives rise to parafollicular cells of thyroid gland).

Pharyngeal Clefts

First cleft forms external auditory meatus. The ectoderm in the floor of first cleft forms otic placode which develops into internal ear. The epithelial lining of the floor of the 1st cleft forms the cuticular layer of tympanic membrane and the cleft becomes the external auditory meatus **(Fig. 4.4)**.

Second cleft onwards: the clefts and arches are covered by the outgrowth of the second arch whichover hangs 2nd to 4th clefts and finally fuses with the outer aspects of the remaining arches. Thus, the space between overhanging second arch and 2nd, 3rd and 4th clefts form **cervical sinus**. In further development, the sinus disappears due to the fusion of the overhanging mesoderm of second arch. Persistent cervical sinus and rupture of pharyngeal membrane leads to communication between pharynx and exterior called **branchial fistula**. The fistula is found along the lateral wall of neck, anterior to sternocleidomastoid muscle. Nonobliteration of cervical sinus results in **cervical/branchial cyst** which is closed at both ends. The cyst may not be visible at birth but later may be seen as an enlargement along the anterior border of sternocleidomastoid muscle. Cysts can be single or multiple and unilateral or bilateral.

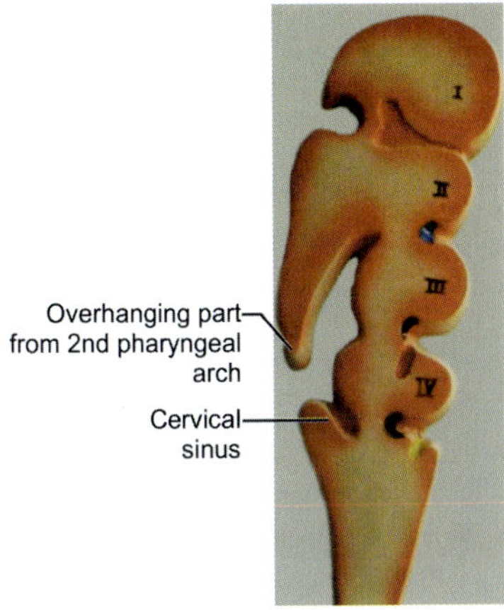

Fig. 4.4: Pharyngeal cleft

DEVELOPMENT OF FACE

There are five primordial or processes involved in the development of face **(Figs 4.5A and B)**.

They are: **frontonasal**, a pair of **maxillary** and a pair of **mandibular** prominences. All these are around the stomodeum (future mouth).

In the frontonasal process (the mesoderm covered by ectoderm lining the forebrain vesicle) on either side of the midline, above the stomodeum, there will be ectodermal thickening called the **olfactory or nasal placode**. Deep to the olfactory placode, there is mesodermal thickening at the circumference of the placode which causes the elevation along the margin of the placode. The raised margins are called the nasal processes; they are **medial nasal**

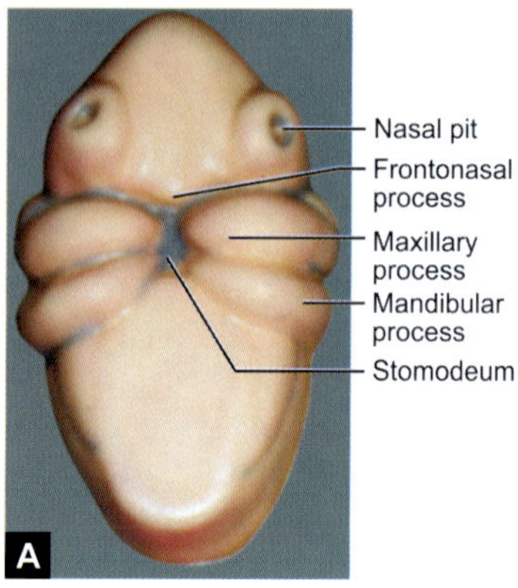

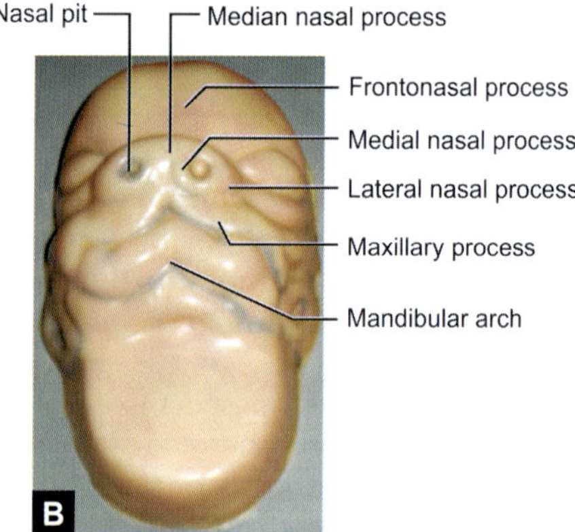

Figs 4.5A and B: Development of face

process and **lateral nasal process.** These nasal processes surround a pit called the **nasal pit**. The nasal pit is floored by olfactory placode. The nasal pit communicates with the stomodeum on the inferior aspect. Maxillary prominences that arise from the cranial part of the first arch grow towards midline compressing medial nasal prominences, thereby the cleft between the prominences is lost and they fuse. The maxillary processes grow towards **median nasal process or globular process or simply the frontonasal process** which is between the two medial nasal processes, fusing to form the cheek and the upper lip.

Hence, formation of upper lip is by fusion of median nasal prominence and maxillary prominences. The median nasal prominence gives rise to **philtrum** of upper lip.

Lower lip and jaw are formed by mandibular prominences as they merge across the midline.

Maxillary and lateral nasal prominences initially are separated by **nasolacrimal groove**.

Ectoderm in the groove forms a solid cord of cells which later gets canalized and forms nasolacrimal duct. Upper end of the duct gets expanded to form **lacrimal sac**.

Nose Development

- Frontonasal prominence—bridge of nose
- Medial nasal prominence—crest and tip of nose and septum of nose
- Lateral nasal prominence—ala of nose.

The intermaxillary segment is composed of:

- Labial component that forms philtrum of upper lip
- Palatal component that forms triangular primary palate.

The muscles of the face (muscles of facial expression) are derived from IInd visceral arch. Therefore, the nerve supply to the muscles of facial expression is by facial nerve and sensory supply is from various branches of trigeminal nerve.

External ears develop on either side of first cleft as auditory hillocks.

Anomalies Associated with the Development of Face

Cleft Lip of Upper Lip

It is caused due to failure of fusion of maxillary process with the frontonasal/median nasal/globular process.

It can be unilateral or bilateral.

Unilateral can be on the right side or left side **(Figs 4.6 and 4.7)**.

This anomaly may be associated with cleft palate.

Hare Lip

Failure of fusion of mandibular processes—it is in the midline **(Fig. 4.8)**.

Oblique Facial Cleft

It is caused due to nonfusion of maxillary process with the lateral nasal process. It is usually associated with absence of nasolacrimal duct.

It can be unilateral or bilateral.

Unilateral can be on the right side or left side **(Fig. 4.9)**.

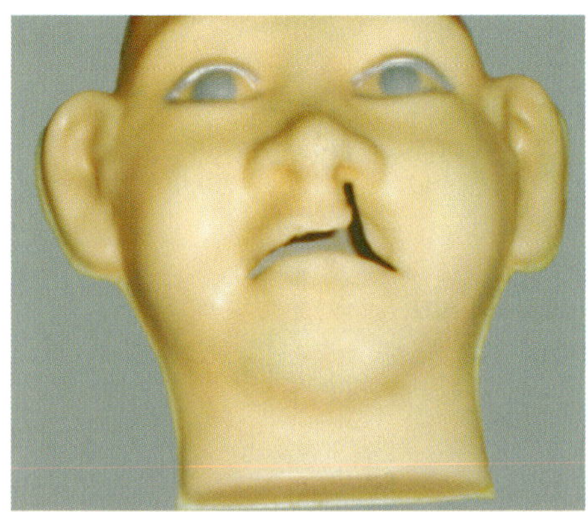

Fig. 4.6: Unilateral cleft lip

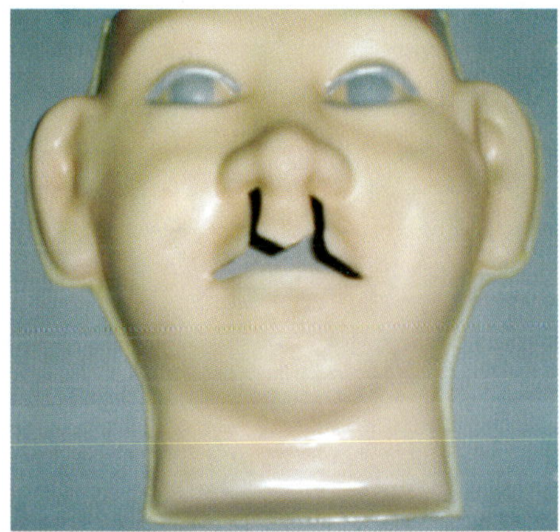

Fig. 4.7: Bilateral cleft lip

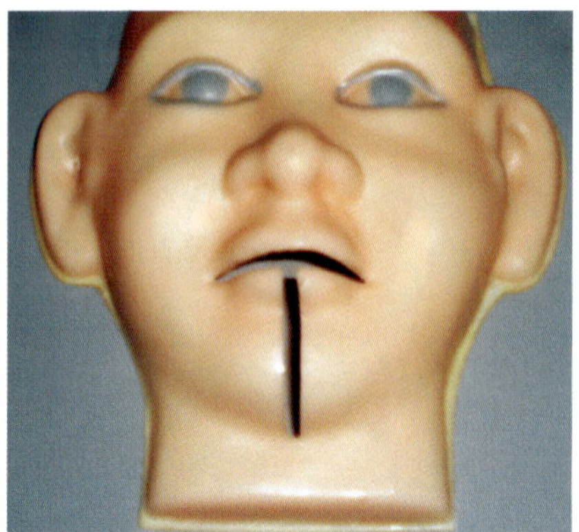

Fig. 4.8: Hare lip

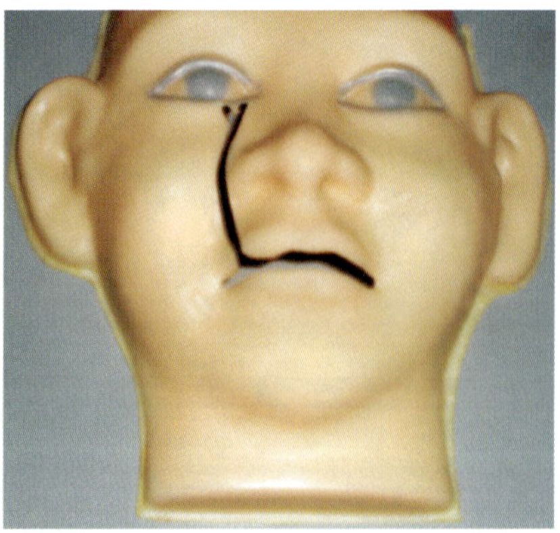

Fig. 4.9: Oblique cleft of the face (unilateral)

DEVELOPMENT OF PALATE (FIG. 4.10)

Primitive or primary palate is formed from intermaxillary segment of frontonasal process. This is adult premaxilla or os incisivum.

Definitive or secondary palate is formed by the fusion of palatal processes or palatal selves of maxillary prominences with the fused primary palate.

The palatal shelves change from a vertical to horizontal position and fuse with each other and with primitive palate, to form secondary palate.

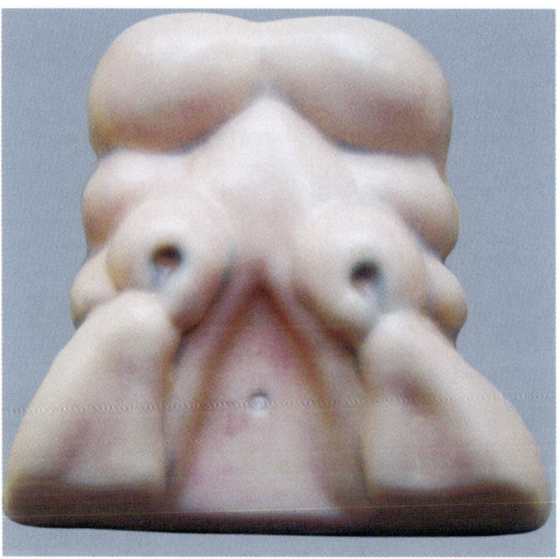

Fig. 4.10: Development of palate

The tongue migrates away from the shelves in an anteroinferior direction for palatal fusion to occur.

The anterior two-thirds undergo ossification and form hard palate.

The posterior parts of the palatine shelves do not ossify and form soft palate.

The nasal pits deepen to form nasal sacs which are separated from the stomodeum by oronasal membrane behind the developing palate. The rupture of oronasal membrane leads to the formation of posterior nasal apertures.

Palatal Anomalies (Fig. 4.11)

Cleft Palate

Anterior cleft palate is lateral to median plane due to the premaxilla in the midline. Therefore, it is either unilateral

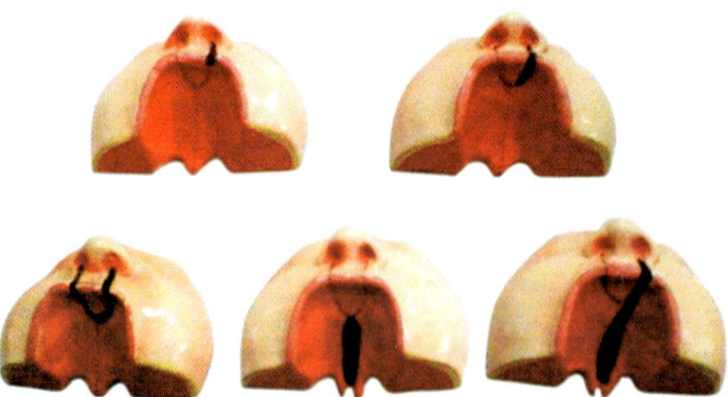

Fig. 4.11: Anomalies of palate

or bilateral. It is caused due to nonunion of premaxilla with the palatine shelf of the maxillary process. It may be associated with cleft lip.

The posterior cleft palate is always in the midline. It can vary in its extent. It may lead to bifid uvula.

There may be nonunion of palatal shelves and the premaxilla leading to complete cleft palate, which is like the letter 'Y'.

DEVELOPMENT OF TONGUE

The tongue develops in the posterior surface of the anterior part of the pharyngeal apparatus.

On the internal surface of first arch, there will be lingual swellings, one on each side of the midline.

Between first and second arches, in the midline there is another swelling called the tuberculum impar.

Caudally, between third and fourth arches, in the midline, another thickening develops called the hypobranchial eminence which consists of cranial and caudal parts.

The lingual swellings grow and fuse with each other.

The cranial part of the hypobranchial eminence **grows** cranially, **over** the tuberculum impar to join with the lingual swellings. The junction or line of fusion is indicated by sulcus terminalis. The tuberculum impar thus gets submerged.

The endoderm covering these swellings give rise to mucous membrane of the tongue. The mesodermal thickenings give rise to connective tissue and muscles of the tongue. The muscles are also developed from occipital myotomes.

Due to the **overgrowth** of the hypobranchial eminence on the second pharyngeal arch to fuse with the lingual swellings, the endoderm covering the third arch spills over and lies in front of the sulcus terminalis in which the circumvallate papillae develop.

Therefore, the nerve supply is correlated to the development as follows:

Presulcal part develops from lingual swellings of first arch, hence general sensory supply is by the lingual branch of mandibular nerve (post-trematic nerve of Ist arch).

Special sensory supply for the taste buds in the fungiform papillae and secretomotor nerves for the lingual glands is by chorda tympani nerve (pretrematic nerve of Ist arch) and for circumvallate papillae, by glossopharyngeal nerve (post-trematic nerve of third arch).

The postsulcal part of the tongue is supplied by glossopharyngeal nerve (post-trematic nerve of third arch) and posterior most part of the tongue is supplied by vagus nerve (contribution from superior laryngeal branch of vagus nerve which is the post-trematic nerve of the fourth arch).

Muscles of the tongue migrate from occipital myotomes, therefore supplied by hypoglossal nerve.

After the tongue is developed, a sulcus, called gingivolingual sulcus appears on either side of the tongue and separates the tongue from the floor of the oral cavity anteriorly leading to freely moving tip and the body of the tongue. The fixed posterior part of the tongue remains as the root of the tongue **(Figs 4.12A and B)**.

DEVELOPMENT OF RESPIRATORY SYSTEM

Consists of:

- Upper respiratory tract
- Lower respiratory tract

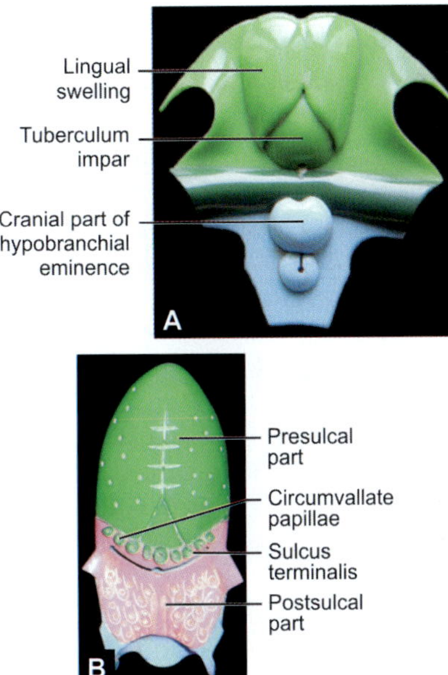

Figs 4.12A and B: Development of tongue

The respiratory system develops after the fourth week of gestation **(Figs 4.13 A to C)**.

Development of Upper Respiratory Tract

Consists of:

- Nose, nasopharynx
- Development of nasal cavity.

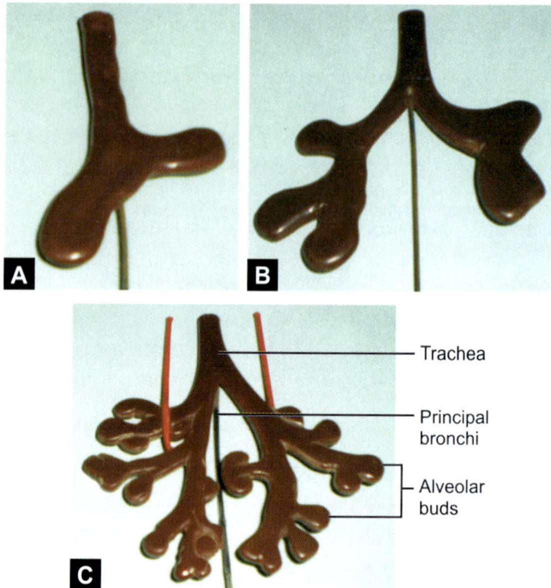

Figs 4.13A to C: Development of respiratory system

Nose

Begins with the formation of:

- Nasal placode
- Nasal pit
- Nasal sac—separated from the primitive oral cavity (stomodeum) by oronasal membrane which consists of ectoderm of stomodeum and ectoderm of nasal sac.

Both nasal sacs are separated by developing nasal septum from posterior part of frontonasal process.

Bucconasal membrane lies behind the primitive palate (premaxilla).

The oro/bucconasal membrane ruptures behind the palate and nasal cavity communicates with oral cavity behind the secondary palate.

This communication is the posterior nasal aperture.

Roof of the nasal cavity is from ectoderm of nasal placode—olfactory epithelium.

Floor—definitive palate.

Lateral wall—lateral nasal process. From this wall, conchae develop to increase the surface area for conditioning the air in the nasal cavity.

Medial wall—frontonasal process—nasal septum.

The pharynx develops from cranial part of the foregut.

The part behind the nasal cavity is called nasopharynx.

Development of Lower Respiratory Tract

Lower respiratory organs include larynx, trachea and bronchus which start developing from fourth week.

Respiratory primordium develops by 28th day as a median outgrowth from the caudal end of ventral wall of primordial pharynx called laryngotracheal groove. Tracheobronchial tree develops caudal to fourth pharyngeal pouch.

Splanchnic mesoderm surrounding the foregut forms connective tissue, cartilages and smooth muscle of respiratory tract.

By the end of 4th week, laryngeal groove evaginates to form respiratory diverticulum (lung bud).

Tracheoesophageal folds appear between tracheobronchial diverticulum and rest of pharynx.

The folds are longitudinal and approach each other to form tracheoesophageal septum which forms partition dividing ventral parts as trachea and dorsal part as esophagus.

Development of Larynx

- Epithelium is derived from the endoderm of cranial end of laryngotracheal tube.
- Laryngeal cartilages are derived from mesenchyme of neural crest incorporated in 4th and 6th arches.
- Arytenoid swellings grow towards tongue.
- Primordial glottis is 'T' shaped laryngeal inlet.
- Epithelium temporarily occludes laryngeal lumen.
- Recanalization starts by 10th week.
  Laryngeal ventricles form during this recanalization.
- Recesses are bounded by folds of mucous membrane called vocal folds and vestibular folds.
- Epiglottis develops from caudal part hypobranchial eminence.
- Laryngeal muscles develop from 4th and 6th pharyngeal arches which are supplied by vagus nerve.
- Growth increases during 1st three years of life.

Development of Trachea

- Endoderm gives rise to epithelium and glands of trachea.

- Splanchnic mesenchyme gives rise to cartilage, connective tissue, and muscles of trachea.

Development of Bronchi and Lungs

- Tracheal buds form two outpouchings, which are called primary bronchial buds.
- These grow laterally into the pericardioperitoneal (future pleural) canals.
- Each divides again into smaller diverticula—two on left and three on right side.
- Segmental bronchi are formed by 7th week.
- Each segmental bronchus and surrounding mass of mesoderm forms primordium bronchopulmonary segment.
- By 24 weeks—about 17 orders of branches have formed and respiratory bronchioles develop.
- Splanchnic mesoderm gives rise to cartilage, smooth muscles, connective tissue and capillaries of trachea.
- Visceral pleura is derived from splanchnic mesoderm.
- Parietal pleura is derived from somatic mesoderm.
- Pleura develop from the walls of pericardioperitoneal canals.

Maturation of Lungs

Divided into four periods:

1. Pseudoglandular period
2. Canalicular period
3. Terminal saccular period
4. Alveolar period.

Pseudoglandular Period

- Lung resembles exocrine gland.
- By 16 weeks, all major elements of lung are formed except for alveoli.
- Fetus is unable to survive at this stage.

Canalicular Period

- It is between 16 and 26 weeks.
- Lumen of terminal bronchi and bronchioles are larger.
- By 24 weeks, each terminal bronchiole has given 2 or more respiratory bronchioles.
- Alveolar ducts are formed from division of respiratory bronchioles.
- At the end of canalicular period, thin walled saccules are formed.
- Lung tissue is well vascularized.

Terminal Saccular Period

- Many alveoli develop.
- Epithelium becomes very thin.
- Capillaries bulge into developing alveoli.
- Intimate contact between epithelial and endothelial cells in the form of blood alveoli barrier appears.

Terminal Alveolar Period

- Type 1 alveolar cells are squamous epithelial cells.
- Type 2 are rounded secretory epithelial cells which secrete pulmonary surfactant, a complex mixture of phospholipids.

- Surfactant reduces surface-tension, forces and facilitates expansion of terminal saccules by preventing their collapse.
- Fetus of less than 24 to 26 weeks has surfactant deficiency.

Alveolar Period

- Type 1 alveolar cells become thin, number of capillaries increase.
- Characteristic mature alveoli do not form until after birth. About 95% of alveoli develop postnataly.

DEVELOPMENT OF HEART

- First major system to function.
- Primordial heart and vascular system starts in the middle of 3rd week of development.
- Heart functioning starts by early 4th week.
- Cardiovascular system develops from splanchnopleuric intraembryonic mesoderm.

Early Development (Figs 4.14A to D)

- Endothelial strands named angioblastic cords appear in cardiogenic mesoderm by 3rd week.
- Cords canalize and heart tubes are formed.
- Heart starts beating by 22nd to 23rd day.
- Blood flow starts by 4th week.
- Primordial heart tube starts developing by 18 days **(Figs 4.14A to D)**.
- Both endothelial lined heart tubes approach and fuse to form single heart tube as a result of lateral folding of embryo.

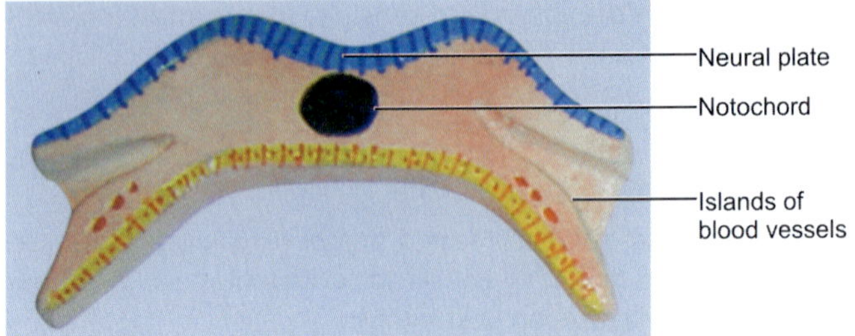

Fig. 4.14A

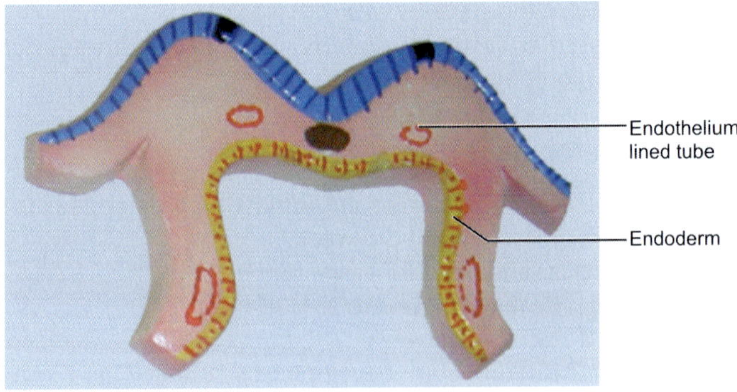

Fig. 4.14B

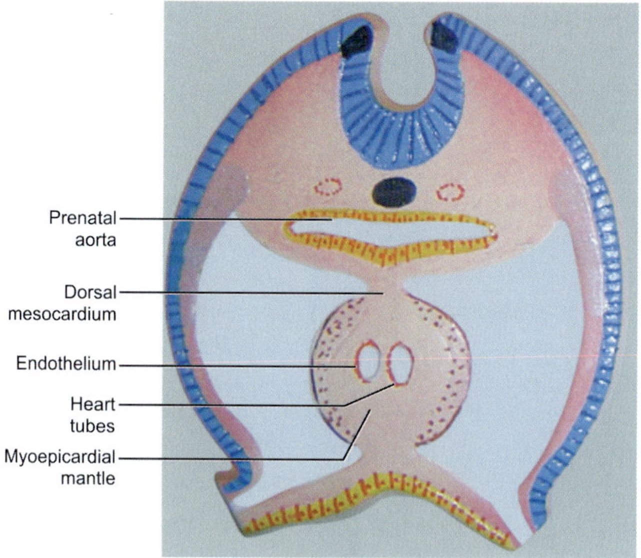

Fig. 4.14C

- Myocardium develops from splanchnic mesoderm called cardiac jelly which is surrounding the pericardial celom.
- Endothelial lining forms endocardium.
- Mesothelial cells form visceral pericardium.
- With the head fold, the heart and pericardial cavity come to lie ventral to foregut and caudal to oropharyngeal membrane.

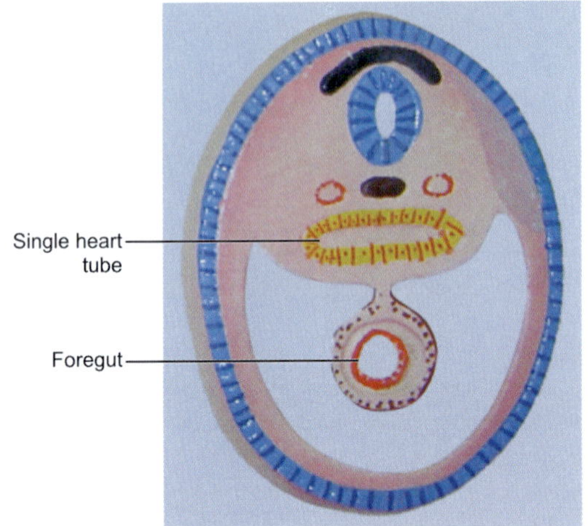

Fig. 4.14D

Figs 4.14A to D: Development of heart tube

Gradually heart tube exhibits alternate dilatation and constrictions craniocaudally. These are:

- Truncus arteriosus
- Bulbus cordis
- Ventricle
- Atrium
- Sinus venosus **(Fig. 4.15)**.

Between dilatations, there are endocardial thickenings at the constriction; between sinus venosus and primitive atrium, they

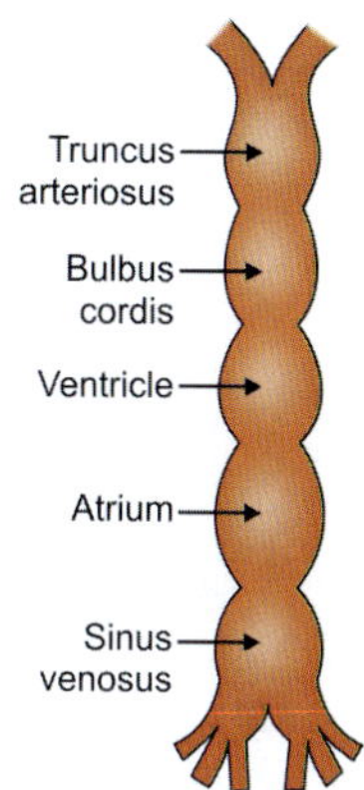

Fig. 4.15: Primitive heart tube

are called sinu atrial valves; between atrium and ventricle, they are termed atrioventricular cushions; between ventricle and bulbus cordis, there are proximal bulbar rides and between bulbus cordis and truncus arteriosus are the distal bulbar ridges.

- Truncus arteriosus is continuous cranially with aortic sac.
- Sinus venosus receives umbilical, vitelline and common cardinal veins from chorion, yolk sac and embryo respectively. These veins open into the corresponding horns of sinus venosus.
- Arterial and venous end of heart tube are fixed.
- Bulbus cordis and ventricle grow faster and form U-shaped tube called bulboventricular loop.
- Heart elongates and bends, gradually invaginates into pericardial cavity.

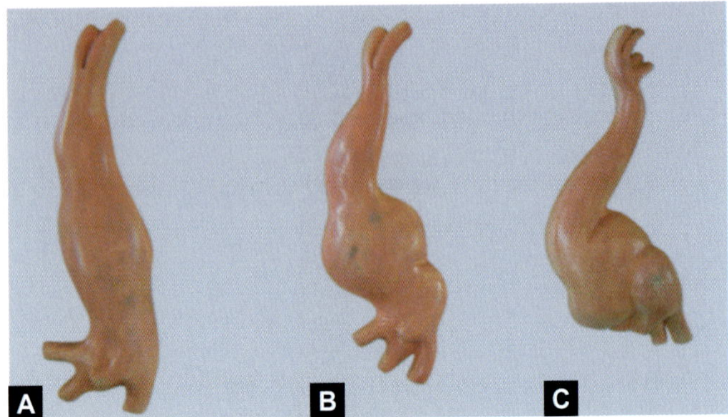

Figs 4.16A to C: Formation of cardiac loop

- Dorsal mesocardium suspends the heart from the anterior wall of the foregut. Gradually, the central part of the dorsal mesocardium disappears forming transverse sinus of pericardium **(Figs 4.16A to C)**.

Partitioning of Primordial Heart

- Partitioning of atrioventricular canal, primitive atrium and primitive ventricle begins around mid 4th week and is completed by 5th week.
- Endocardial cushions appear from dorsal and ventral wall of AV canal and are formed from cardiac jelly.
- Endocardial cushions separate primitive atrium and primitive ventricle and function as AV valves.

- AV cushions approach each other and fuse to form intermediate AV cushion.
- Formation of AV cushions results in right and left AV canals.
- Endocardial cushions separate atria and ventricle and function as AV valves.
- Endocardial cushions are formed by specialized extracellular matrix or cardiac jelly.

Partitioning of Primordial Atrium

- Begins at the end of 4th week.
- Septum primum, a thin membrane, descends from the roof of primary atrium to intermediate AV cushion.
- Before this fusion, there is a communication between right and left halves of primitive atrium. It is called foramen primum **(Figs 4.17A to D)**.
- Foramen primum acts as a shunt for oxygenated blood to pass from right to left atrium.
- Septum primum fuses with intermediate endocardial cushion to form primordial interatrial septum.
- Before foramen primum disappears, perforations appear in the central part of septum primum.
- Perforations coalesce to form foramen secundum. Foramen primum obliterates.
- A crescent shaped septum secundum grows from the roof of primitive atrium to the right of septum primum partially closing the foramen primum and converting it into foramen ovale.
- Before birth, foramen ovale transmits oxygenated blood from right to left atria. Left to right is prevented by septum primum closing on septum secundum.

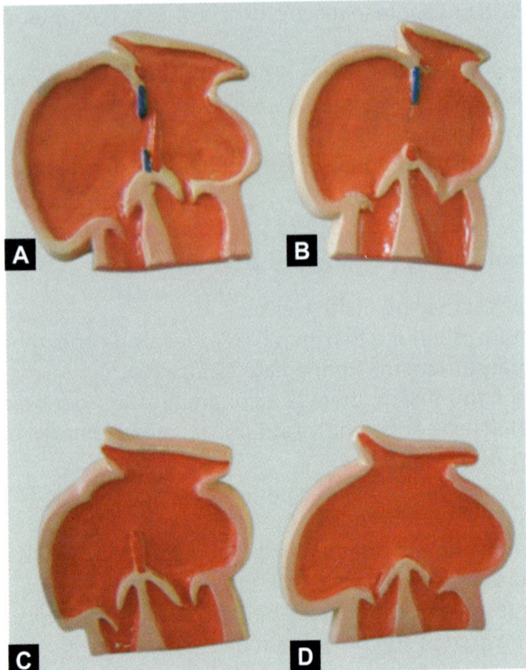

Figs 4.17A to D: Development of interatrial septum

- After birth, foramen ovale closes due to increased pressure in the left atrium after the establishment of pulmonary circulation.
- Valve of foramen ovale fuses with septum secundum thus obliterating the foramen ovale.

- Oval depression in the lower part of interatrial septum of right atriums a remnant of septum primum and is called fossa ovalis and its margin called limbus fossa ovalis is the remnant of septum secundum fused with septum primum.

Partitioning of Primordial Ventricle

- It is first indicated by a median muscular ridge in the floor of the ventricle near the apex.
- Crescentic fold with concave free edge is formed which forms muscular part of interventricular septum.
- Increase in the height of muscular part of septum results from dilatation of ventricles on each side of interventricular septum.
- Until 7th week, septum is, crescent shaped between free edge of interventricular septum and intermediate endocardial cushion.
- Communication between right and left ventricle is through interventricular foramen which closes by the end of 7th week as bulbar ridges fuse with intermediate endocardial cushion **(Figs 4.18A to C)**.
- Formation of the membranous part of interventricular septum followed by closure of interventricular foramen results from fusion of tissues from three sources.
  1. The right bulbar ridge
  2. The left bulbar ridge
  3. The intermediate endocardial cushion.
- After closure of interventricular foramen and formation of membranous part of interventricular septum, pulmonary trunk communicates with right ventricle, ascending aorta communicates with left ventricle.

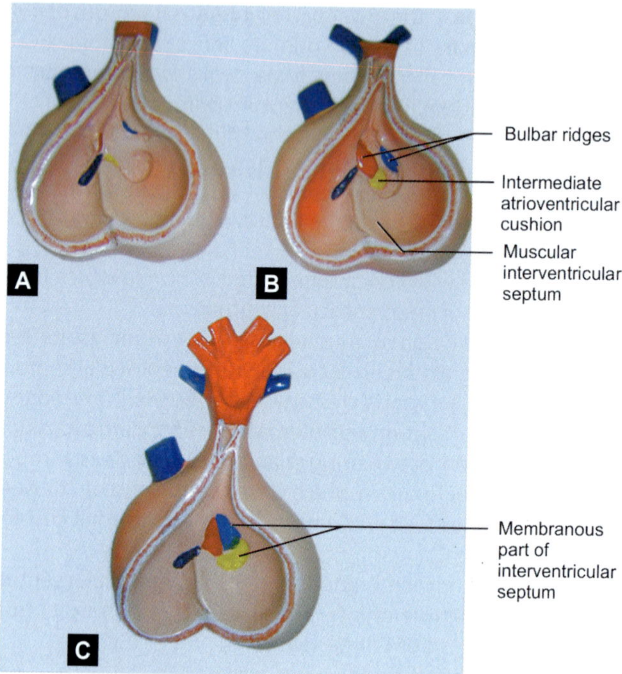

Figs 4.18A to C: Development of interventricular septum

- Cavitations of ventricular valves in the form of sponge work of muscular bundle develop into trabeculae carneae.
- Some bundles become the papillary muscles and tendinous cords (chordae tendinae) extending between papillary muscles to AV valves.

AORTIC ARCHES AND DERIVATIVES (FIGS 4.19 TO 4.22)

- Pharyngeal arches develop during 4th week.
- Primitive aortae develop on either side of developing vertebral column.
- They arch forwards to communicate with primitive heart.
- The arching takes place in the region of first pharyngeal arch.
- The pats of the primitive aortae on either sides of the vertebral column are called **dorsal aortae**.

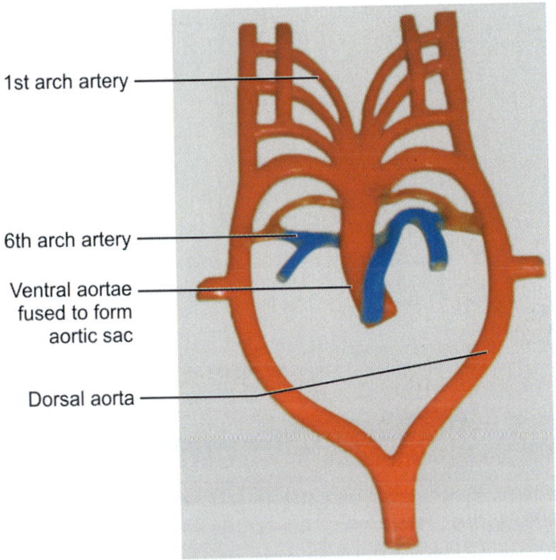

Fig. 4.19: Development of aortic arches

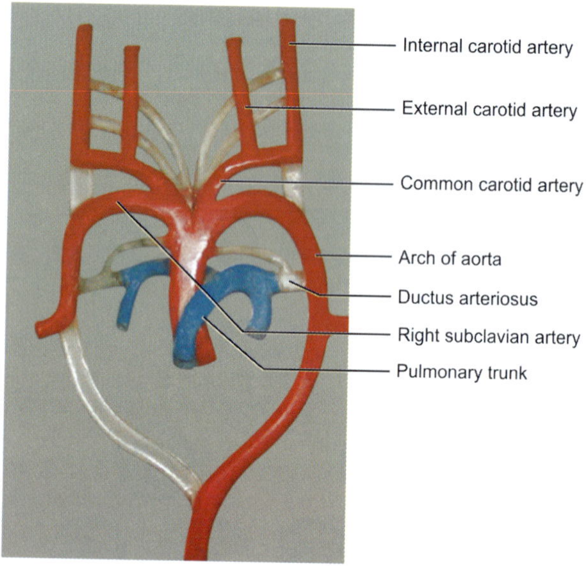

Internal carotid artery
External carotid artery
Common carotid artery
Arch of aorta
Ductus arteriosus
Right subclavian artery
Pulmonary trunk

Fig. 4.20: Derivatives of aortic arches

Axis artery of upper limb
Median artery

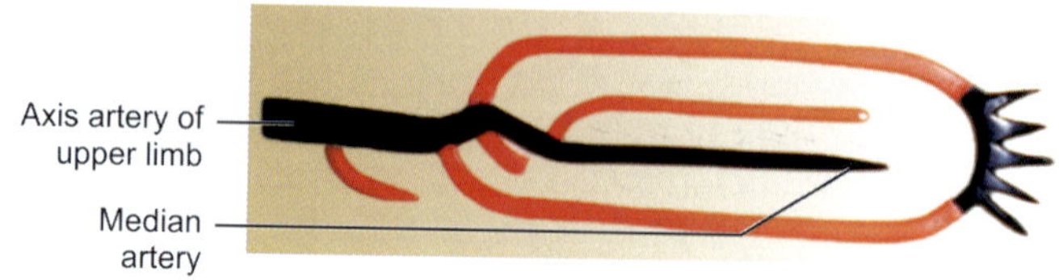

Fig. 4.21: Axis artery of upper limb

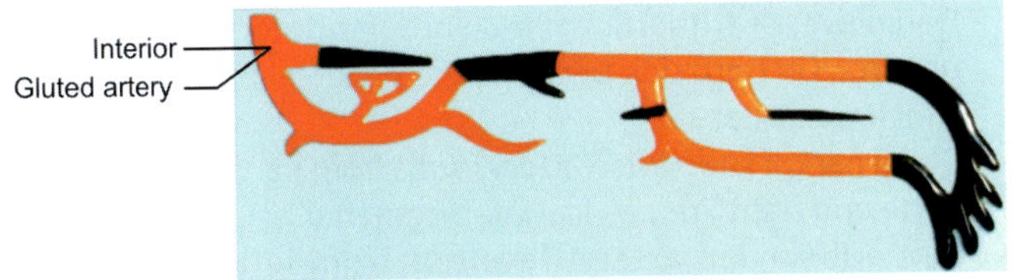

Fig. 4.22: Axis artery of lower limb

- The primitive aortae which are extending anteriorly are called **ventral aortae**.
- The ventral and dorsal aortae on each side are connected by an arch.
- The arched part in the region of first pharyngeal arch is the **first aortic arch** or 1st pharyngeal arch artery.
- As the succeeding pharyngeal arches appear there will be anastomosis between dorsal and ventral aortae of that side lying on the corresponding sides of pharyngeal arches.
- Six pairs are formed in which first two disappear when 6th pair of arteries appear.
- During 8th week, these are transformed into the final fetal arterial arrangement.
- Dorsal parts of first aortic arches form maxillary arteries.
- Derivatives of second pair of aortic arch arteries in dorsal part persist as stems of stapedial arteries.
- Derivatives of 3rd pair of aortic arches in the proximal part form common carotid arteries and distal parts join with dorsal aortae and form internal carotid arteries.
- External carotid arteries appear as buds from third arch arteries.

- Derivatives of 4th aortic arches—on left side forms part of arch of aorta. On right side becomes the proximal part of the right subclavian artery.
- Distal part of subclavian artery forms from part of right dorsal aorta and right 7th cervical intersegmental artery.
- Left subclavian artery develops from left 7th cervical intersegmental artery.
- Because of differential growth, the left subclavian artery comes to lie close left common carotid artery.
- Fate of 5th pair of aortic arch artery is, in 50% of embryos 5th pair is rudimentary and soon degenerates and in other 50% they do not develop.
- 6th pair of aortic arches on left side—proximal part persists as proximal part of left pulmonary artery and distal part of the arch artery passes from left pulmonary artery to dorsal aorta to form arterial shunt called ductus arteriosus.
- Right 6th arch artery—proximal part of it persists as proximal part of right pulmonary artery and distal part degenerates.
- Transformation of 6th pair of aortic arches determines the course of recurrent laryngeal nerves.
- On the right side as distal part of right 6th aortic arch degenerates right recurrent laryngeal nerve hooks around right subclavian artery—derivative of 4th aortic arch.
- On the left side, the recurrent laryngeal nerve hooks around the ductus arteriosus, formed by distal part of 6th aortic arch.
- When arterial shunt involutes after birth, the nerve hooks around the ligamentum arteriosum and the arch of the aorta.

FETAL CIRCULATION (FIG. 4.23)

Oxygen and nutrient rich blood from placenta through left umbilical vein enters ductus venosus, which connects left umbilical vein to inferior vena cava.

Most of the blood from left umbilical vein and vitelline veins bypasses the liver and the remaining portion of blood flows into sinusoids of the liver. It then enters inferior vena cava through hepatic veins and finally enters right atrium. Through foramen ovale blood enters left atrium and right ventricle.

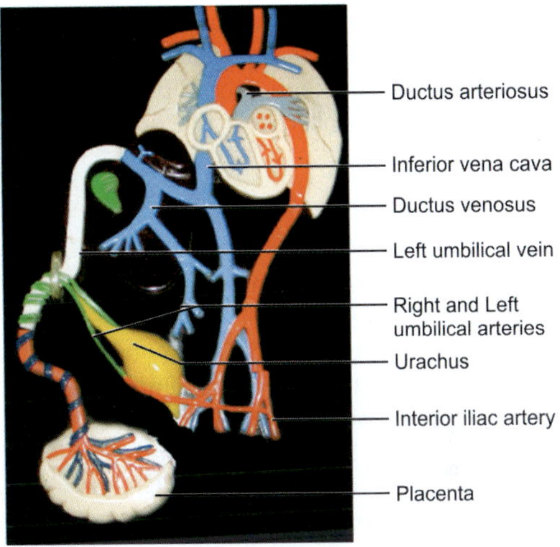

Fig. 4.23: Fetal circulation

From left atrium, most of the blood goes to left ventricle from where it reaches aorta, remaining blood goes to the lungs. The blood from right ventricle goes to pulmonary trunk. From there, the blood enters into ductus arteriosus.

Through ductus arteriosus, blood enters descending aorta.

From here, the blood is returned to the placenta for reoxygenation via right and left umbilical arteries which are the branches of internal iliac arteries.

Adult derivatives of fetal vascular structures:

- Left umbilical vein—ligamentum teres of liver
- Ductus venosus—ligamentum venosum
- Umbilical arteries—proximally become superior vesical arteries and distally medial umbilical ligaments
- Ductus arteriosus—ligamentum arteriosum
- Intra-abdominal part of allantois—urachus (Median umbilical ligament).

Portal Vein (Fig. 4.24A)

There are three pairs of veins opening into the developing heart (sinus venosus). These are:

1. Common cardinal veins (Ducts of Cuvier), formed by the fusion of anterior cardinal and posterior cardinal veins which drain the deoxygenated blood from body wall.
2. Vitelline veins—drain the deoxygenated blood from gut and yolk sac.
3. Umbilical veins—drain oxygenated blood from placenta.

Vitelline Veins (Fig. 4.24A)

- Vitelline veins start in the yolk sac.
- By 12 weeks, the left vitelline vein has regressed and the blood from the left side of the abdominal viscera drains to the right.
- The enlarged right vitelline vein between septum transversum and sinus venosus is called right hepatocardiac channel.
- Sinusoids in the liver derived from the vitelline veins and umbilical veins form a final channel which drains into the inferior vena cava.

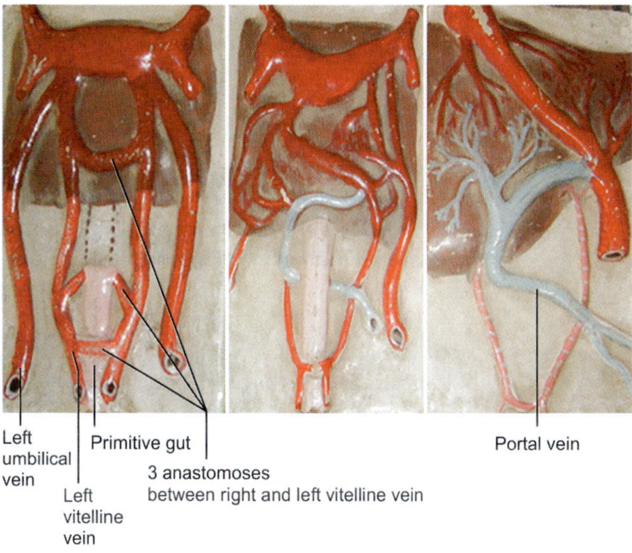

Fig. 4.24A: Development of portal vein

- Right umbilical vein becomes obliterated during the 2nd month of intrauterine life due to the kinking action of developing liver.
- Left umbilical vein anastomoses with the ductus venosus. It is a shunt between left umbilical vein and right hepatocardiac channel (which contributes in the formation of inferior vena cava).

Additionally, the development of the liver from the mesenchymal cells of the septum transversum imposes vascular changes in the process of forming the hepatic sinusoids.

There will be anastomosis between right and left vitelline veins at three levels around the developing duodenum. These anastomoses are—cranial ventral, dorsal and caudal ventral.

The persisting parts of the vitelline veins and the anastomoses give rise to portal vein. They are:

Ventral cranial anastomosis—forms left branch of portal vein.

Right vitelline vein superior to ventral cranial anastomosis—forms right branch of portal vein.

Right vitelline vein between ventral cranial and dorsal anastomosis—forms supraduodenal part of portal vein.

Dorsal anastomosis forms the retroduodenal part of the portal vein.

Cranial part of left vitelline vein between dorsal anastomosis and caudal ventral anastomosis—forms infraduodenal part of portal vein.

Superior mesenteric vein and splenic vein which drain separately into infraduodenal part of left vitelline vein get absorbed into the developing portal vein.

Inferior Vena Cava (Fig. 4.24B)

Components

- Hepatic segment—from right vitelline vein (right hepatocardiac channel).
- Subhepatic segment—from anastomosis of right hepatocardiac channel and right subcardinal vein.
- Prerenal segment—from right subcardinal vein.
- Renal segment—from subcardinal and supracardinal anastomosis.

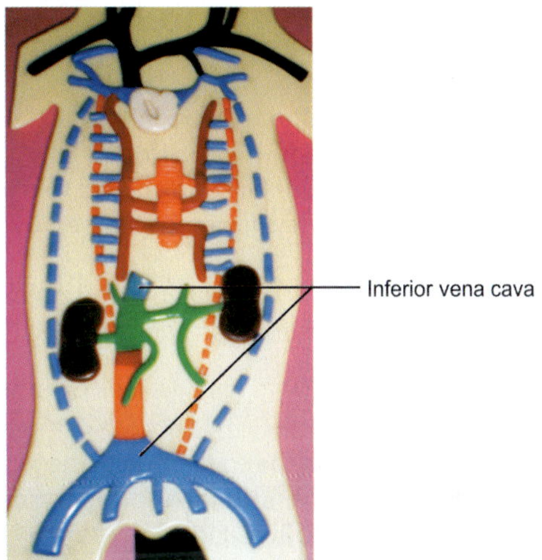

Fig. 4.24B: Development of inferior vena cava

- Postrenal segment—from right supracardinal vein.
- Postcardinal—supracardinal anastomosis.
- Iliac segment—from right posterior cardinal vein.

GUT DEVELOPMENT

- The gut tube develops from endoderm, visceral mesoderm and celomic epithelium including part of yolk sac **(Figs 4.25A and B)**.
- The boundaries between the three parts of the developing gut are indicated by the margins of distribution of the primary arteries of the three parts.
- The extensive elongation results in folding and rotation of developing gut tube.
- The tube develops through a stage in which the lumen is obliterated and subsequently recanalized.
- Where layers of peritoneum become applied to one another due to folding, the deep layers are reabsorbed by a process called zygosis.
- Abnormalities of the system are related to the major events of elongation, recanalization, rotation and septation.

Foregut

- The foregut develops from the cranial part of the gut. Includes the pharynx, esophagus, stomach, duodenum and the associated salivary glands, liver and pancreas.
- It receives its innervation from the glossopharyngeal and vagus nerves and from the cranial and thoracic sympathetic chain.

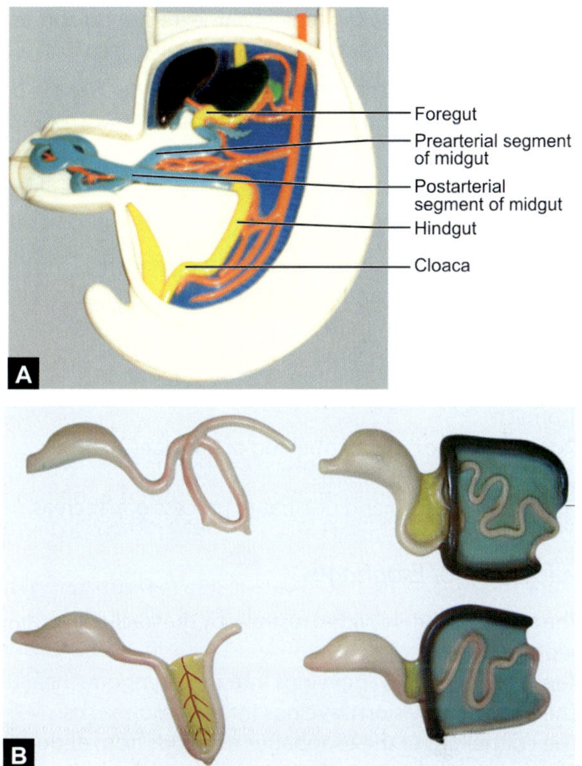

Figs 4.25A and B: Development of gut

- Blood supply is by branches of the external carotid artery, thyrocervical trunk, thoracic aorta and celiac trunk.

Venous drainage of the intra-abdominal foregut is to the portal venous system, with an anastomosis around the lower esophagus between portal and systemic venous systems.

Derivatives of Foregut

- Oral cavity, tongue, tonsils and salivary glands
- Primordial pharynx and derivatives
- Upper respiratory system
- Lower respiratory system
- Esophagus
- Stomach
- Duodenum proximal to opening of bile duct
- Liver and biliary apparatus
- Lower part of head and uncinated process of pancreas.

Development of Esophagus

- The foregut tube is divided rostrally by the tracheoesophageal septum.
- The anterior division develops into the laryngotracheal tube.
- The posterior division develops into the esophagus.
- The epithelium of the esophagus develops from endoderm.

Proliferation of the epithelium commonly obliterates the lumen (squamous cell metaplasia) but further growth results in recanalization by the end of the embryonic period. The muscle coat of the esophagus is derived from intraembryonic visceral mesoderm.

Development of Stomach

- The caudal part of the foregut develops into the stomach.
- Initially the primordium of the stomach lies in the midline.
- There will be fusiform dilatation.
- Differential growth of the fusiform dilatation takes place, being faster along the dorsal border than the ventral.
- As this process is occurring, the tube rotates for about 90°—anterior border to the right side and posterior border to the left side.
- The end result of this differential growth is that the cranial end of the stomach lies to the left of the midline, the body lies almost at right angles to the midline, and the pylorus lies to the right of the midline.
- The fast growing part becomes greater curvature and slow growing part appears drawn in and forms lesser curvature.
- The growth of the stomach causes the ventral mesentery running from the liver to the lesser curvature to lie almost vertically as the lesser omentum. The dorsal mesentery expands to drape over the small bowel and run back to attach to the posterior abdominal wall behind the stomach to form greater omentum.

Development of Duodenum

Duodenum develops from caudal most part of the foregut and cranial part of prearterial segment of midgut. The changes that occur are:

- Rotation of gut 90°.
- Differential growth—right wall grows faster and forms 'C'-shaped curve.

- Zygosis—fusion and absorption of peritoneum on posterior part, pushing it retroperitoneally.

At the junction between foregut and midgut there is opening of hepatopancreatic duct.

Development of Midgut

The midgut forms a U-shaped loop that herniates into the umbilical cord during the 6th week of gestation (physiological umbilical herniation).

While in the umbilical cord, the midgut loop rotates 90°. Artery of the midgut is superior mesenteric artery and forms axis of midgut loop.

During the 10th week of gestation, the midgut loop returns to the abdomen, rotating to an additional 180°. Thus, gut consists of prearterial limb and postarterial limb.

The cranial limb (prearterial segment) of the midgut elongates rapidly during development and forms the jejunum and cranial portion of the ileum.

The caudal limb (postarterial segment) forms the cecum **(Figs 4.26A and B)**, appendix, caudal portion of the ileum, ascending colon and proximal two-thirds of the transverse colon.

The caudal limb is easily recognized during development because of the presence of the cecal diverticulum.

The midgut loop rotates to 270° counterclockwise around the superior mesenteric artery as it retracts into the abdominal cavity during the 10th week of development.

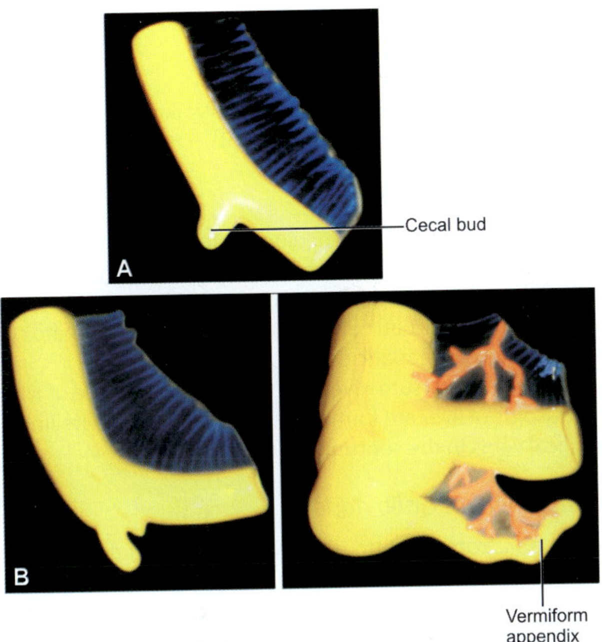

Figs 4.26A and B: Development of cecum

Midgut Derivatives

- Distal duodenum—distal to the opening of hepatopancreatic duct
- Jejunum and ileum

- Vermiform appendix
- Ascending colon
- Right two-thirds of transverse colon
- The midgut communicates with the yolk sac via the vitello-intestinal duct (yolk stalk).

Development of Liver

It develops as hepatic diverticulum from the caudal most part of the foregut. It arises from the anterior wall of the endoderm of the foregut in the ventral mesogastrium, towards the septum transversum. The hepatic diverticulum gives rise to hepatocytes, bile canaliculi and hepatic ducts.

Septum transversum gives rise to connective tissue of liver.

Sinusoids of liver are formed by the branching of vitelline and umbilical veins in the septum transversum.

Development of Gallbladder

It develops from hepatic diverticulum as pars cystica which becomes gallbladder and cystic duct.

Development of Pancreas

Develops from endoderm as ventral and dorsal pancreatic buds arising from anterior wall of midgut and posterior wall of foregut respectively at the junction between foregut and midgut.

Due to 90° of rotation of foregut, differential growth of duodenal wall and absorption of peritoneum (zygosis) on the dorsal part of the duodenum, the ventral pancreatic bud moves to the left where the dorsal pancreatic bud is situated. Both buds fuse and form pancreas. The ventral bud gives rise to lower part of head and uncinate process; the dorsal bud gives rise to upper part of head, neck, body and tail of pancreas.

Exocrine part: Development of duct system—main pancreatic duct develops from:

- The duct of the ventral bud
- Communication between ducts of ventral and dorsal buds
- The distal part of the duct of the dorsal bud.

Accessory pancreatic duct develops from:

- The proximal part of the duct of the dorsal bud
- The end pieces of the terminal ducts expand to form acini.

Endocrine part of pancreas: Before canalization of the ducts and formation of acini of exocrine part, clusters of endodermal cells get detached and form islets of Langerhans.

Development of Hindgut

Cloaca (Figs 4.27 and 4.28)

The caudal dilated part of the hindgut is known as the cloaca. At the end of the cloaca, there is a cloacal membrane which separates the endodermal cloaca from the surface ectodermal proctodeum (a depression around the cloacal membrane). Cloacal membrane is formed by endoderm and ectoderm.

Partitioning of cloaca: The cloaca is the endodermally lined cavity at the end of the gut tube.

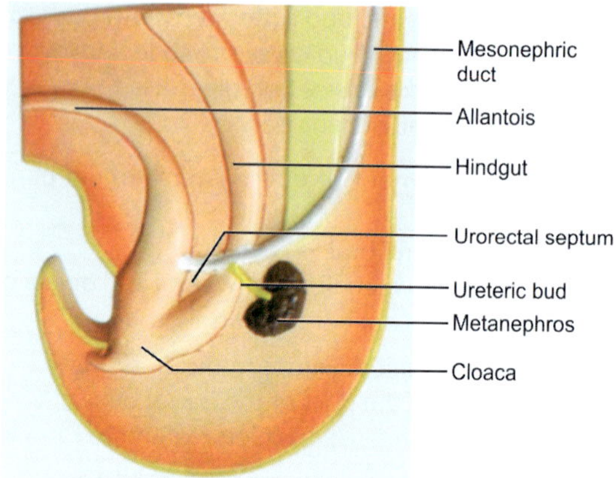

Fig. 4.27: Cloaca

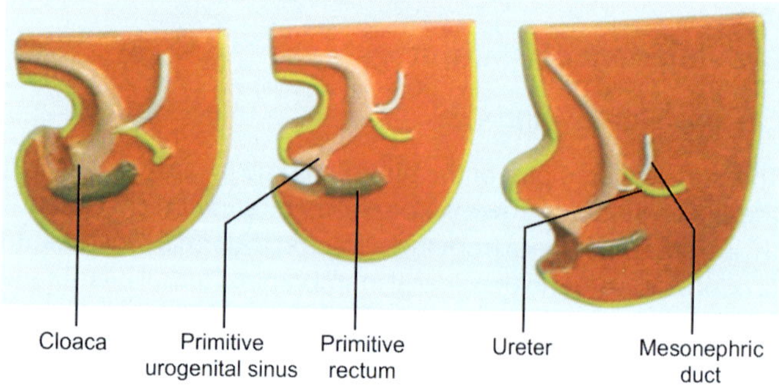

Fig. 4.28: Derivatives of cloaca

It has a diverticulum into the body stalk called the **allantois,** which is a diverticulum from the caudal wall of the yolk sac.

- The **cloacal membrane** separates the cloaca from the proctodeum (**anal pit**).
- During development a sheet of mesoderm (**urorectal septum and lateral mesodermal folds**) develop between allantois and hindgut, to divide the cloaca into a ventral (**urogenital sinus**) and a dorsal portion (**anorectal canal**).
- By 7th week, the urorectal septum reaches the cloacal membrane, dividing it into ventral **urogenital membrane** and dorsal **anal membrane**.
- The epithelium of the superior two-thirds of the anal canal is derived from the endodermal hindgut.
- The inferior one-third develops from the proctodeal ectoderm.
- The junction of these two epithelia is indicated by the **pectinate line**, which also indicates the former site of the **anal membrane** that normally ruptures during the **8th week** of development.

Hindgut Derivatives

- Left one-third of transverse colon
- Descending colon, sigmoid colon, and rectum
- Proximal anal canal (superior to the pectinate line).

The caudal part of the hindgut known as the cloaca is divided by the urorectal septum into anterior urogenital sinus, which gives rise to urogenital system and posterior primitive rectum that forms the rectum and proximal anal canal.

UROGENITAL SINUS (TABLE 4.2)

Table 4.2: Summary of urogenital system

Embryonic structure	*Male*	*Female*
Indifferent gonad	Testis	Ovary
Cortex	Seminiferous tubules	Ovarian follicles
Medulla	Rete testis	Rete ovarii
Gubernaculum	Gubernaculum testis	Ovarian and round ligament of uterus
Mesonephric tubules	Ductuli efferentes	Epoophoron, paroophoron
Mesonephric duct	Appendix of epididymis, duct of epididymis, ductus deferens, ureter, calices and collecting tubules, ejaculatory duct and seminal vesicles	Appendix vesiculosa, duct of epoophoron, duct of Gartner, ureter, pelvis, calices and collecting tubules
Paramesonephric duct	Appendix of testis	Hydatid of Morgagni, uterine tube, uterus
Urogenital sinus	Urinary bladder, urethra, prostatic utricle, prostate gland and bulbourethral glands	Urinary bladder, urethra, vagina, urethral, paraurethral and greater vestibular glands
Sinus tubercle	Seminal colliculus	Hymen
Phallus	Penis	Clitoris
Urogenital folds	Ventral aspect of penis	Labia minora
Labioscrotal swellings	Scrotum	Labia majora

Formation and Fate

It is the anterior part of the cloaca.

The mesonephric ducts open into the posterolateral parts of the primitive urogenital sinus and dividing it into the upper vesicourethral canal and lower definitive urogenital sinus. The definitive urogenital sinus is further divided into cranial pelvic part and caudal phallic part.

The vesicourethral canal gives rise to:

- Urinary bladder except trigone in both sexes. Trigone develops from mesonephric ducts.
- Anterior wall and upper part of the posterior wall of the prostatic urethra in male.
- Entire female urethra.

Pelvic part of urogenital sinus gives rise to:

- Lower part of posterior wall of prostatic urethra and membranous urethra in male and bulbourethral glands.

Phallic part:

- Most of the spongy urethra in male
- Vestibule in female and greater vestibular glands.

URINARY SYSTEM

From intermediate mesoderm—visible at 10 somite stage.

In lower vertebrates—serial segmental diverticuli appear called Nephrotomes.

Each nephrotome has a cavity—nephrocele which communicates with peritoneal funnel called Nephrostome.

Dorsal wall of nephrotome evaginates as a nephric tubule which bends caudally at the dorsal tips and fuse with the lower ones to form a longitudinal primary—excretory duct.

This duct curves ventrally and opens into Cloaca.

Glomeruli arise from either ventral wall of nephrocele or roof of celom adjacent to the peritoneal funnels or from both.

Renal excretory system is regarded as having 3 stages in development of urinary system—Pronephros, Mesonephros, and Metanephros (in higher vertebrates).

Pronephros

Present as clusters of cells in cranial part.

Caudally-similar groups become vesicular.

Peritoneal funnels are rudimentary and glomeruli are internal and external but urine that is formed drained into peritoneal funnel.

Primary excretory duct: A solid rod of cells undergo canalization in the dorsal part of nephrogenic cord **(Fig. 4.29)**. Cranial end is at 9th somite level and caudal end merges in undifferentiated mesenchyme of cord. The duct later elongates and its caudal end becomes detached from the nephrogenic cord. From this level it grows caudally—independent of nephrogenic mesoderm and then curves ventrally to reach the wall of cloaca.

Mesonephros

Extent: From septum transversum to 3rd lumbar segment:

The intermediate mesoderm now develops into mesonephric tubule which begins to open into excretory duct—now called Mesonephric duct (Wolffian duct).

Mesonephric tubules have been estimated to be 70 to 80 and gradually glomeruli develop at one end where there is dilatation. The other end of tubule opens into mesonephric duct. The numbers of tubules which are present are not more than 30 to 40 at any given time because when the caudal tubules are developing the cranial ones atrophy.

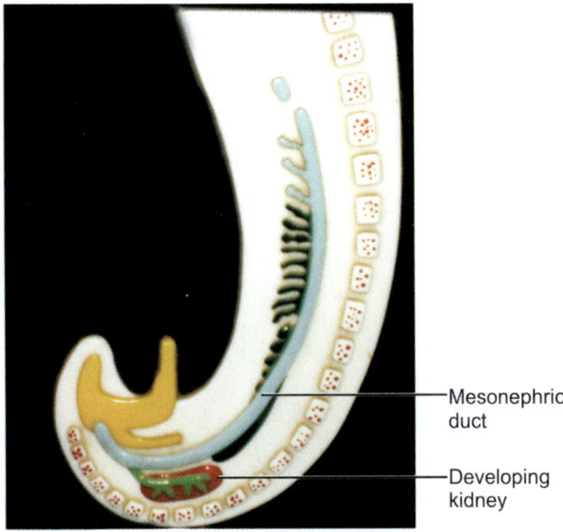

Fig. 4.29: Formation and fate of nephrogenic cord

The whole of the mesonephric tubule now becomes elongated and spindle shaped and projects into celomic cavity lateral to dorsal mesentery, in lower thoracic and lumbar region. It is called mesonephric ridge/Wolffian body. Along with the celomic mesothelium, it is called the genital ridge, which gives rise to gonads **(Figs 4.30 to 4.32)**.

The mesonephric ducts open into cloaca. Just before opening into cloaca, the mesonephric duct gives rise to ureteric diverticulum.

It grows towards metanephros to form the ureter, pelvis of kidney, major and minor calyces.

Fig. 4.30: Development of gonads

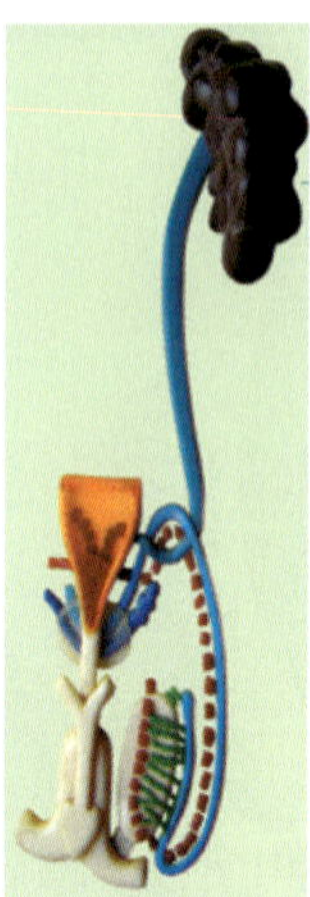

Fig. 4.31: Development of testis

Fig. 4.32: Development of ovary

Metanephros

It is the permanent kidney, appearing in 5th week of intrauterine life. Excretory tubules appear, lengthen rapidly to form an S-shaped loop and acquire a tuft of capillaries that will form the glomeruli. Around the glomerulus, excretory tubules form the Bowman's capsule and the rest of the nephron **(Fig. 4.33B)**. Nephrons communicate with the terminal branches of ureteric bud derived from mesonephric duct.

Thus, kidney develops from **(Figs 4.33A and B)**:

- Metanephric blastema that gives rise to secretory part of the kidney.

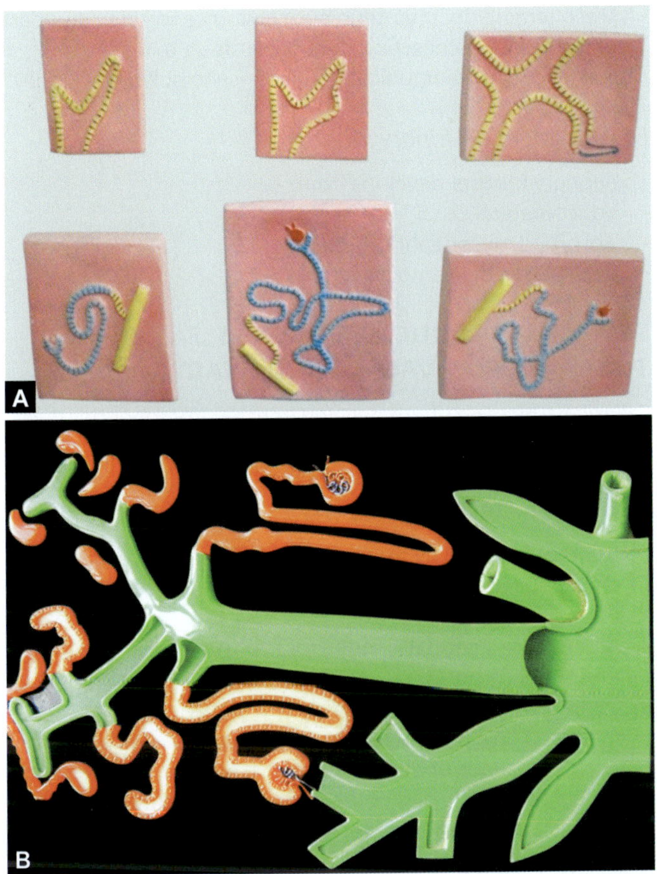

Figs 4.33A and B: Development of kidney

- Mesonephric duct (ureteric bud) gives rise to collecting part of the kidney—collecting duct onwards up to ureter. Failure of this communication gives rise to polycystic kidney.

Development of Urinary Bladder

The urinary bladder develops from:
- Vesicoureteral canal
- Absorbed mesonephric ducts
- A small part of allantois.

FEMALE REPRODUCTIVE ORGANS: FALLOPIAN TUBES, UTERUS, VAGINA AND OVARY

Fallopian tubes and uterus develop from paramesonephric ducts.

Paramesonephric Ducts (Müllerian Ducts Fig. 4.34A)

- Ducts are formed by the invagination of celomic epithelium.
- They lie lateral to the mesonephric ducts in the cranial part of the nephrogenic cord.
- When traced caudally they cross to the medial side of the mesonephric ducts.
- In the midline, the right and the left paramesonephric ducts meet and fuse to form uterovaginal canal.
- The caudal ends of this canal come in contact with the dorsal wall of the phallic part of the definitive urogenital sinus.
- This part of the sinus gives rise to vestibule.
- The paramesonephric ducts develop in the female because of a lack of anti-Müllerian hormone.

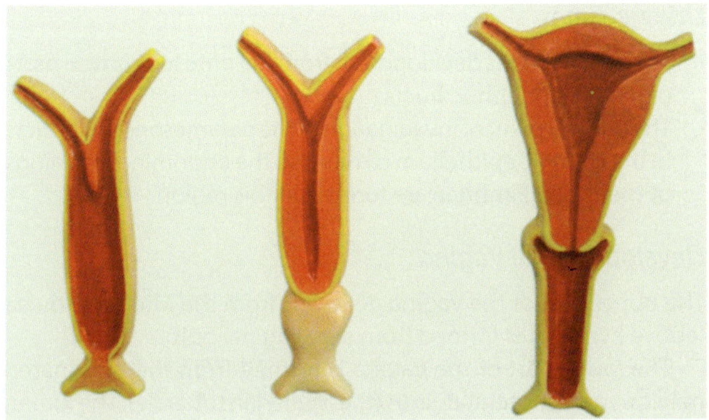

Fig. 4.34A: Development of female reproductive organs

- In the male, it degenerates, leaving vestigial structures such as the appendix of testis from cranial end and prostatic utricle from caudal fused ends.

Development of Uterus

- The epithelium of the uterus develops from the fused paramesonephric ducts—uterovaginal canal.
- The myometrium is derived from surrounding mesoderm.
- As the thickness of the myometrium increases, unfused horizontal parts of two paramesonephric ducts partially get embedded within its substance to form the fundus.
- The cervix can be recognized as a separate region.
- In the fetus, the cervical part is longer than the body of uterus.

Uterine Tubes

- The uterine tubes development from the unfused vertical parts of paramesonephric ducts.
- The original point of invagination of the paramesonephric ducts in the celomic epithelium remains as the abdominal openings of the tubes. Fimbriae are formed in this region.

Development of Vagina

The upper part of the vagina develops from the Müllerian ducts (above hymen it is formed from sinu vaginal bulb).

The lower part of the vagina is formed from the evaginated pelvic part of urogenital sinus (the Müllerian tubercle). The lumen of the vagina remains separated from that of the urogenital sinus by a thin tissue plate called the hymen.

Thus, vagina develops from:

- Above hymen—upper four fifth—the mucosa is derived from sinovaginal bulbs. The musculature develops from the mesoderm of lower fused parts of paramesonephric ducts.
- Below the hymen—lower one fifth—develops from endoderm of urogenital sinus.
- External vaginal orifice—ectoderm of genital folds after the rupture of urogenital membranes.

Uterus Didelphys (Fig. 4.34B)

It is duplication of uterus resulting from lack of fusion of paramesonephric ducts throughout their normal line of fusion leading to uterus didelphys **(Fig. 4.34B)**.

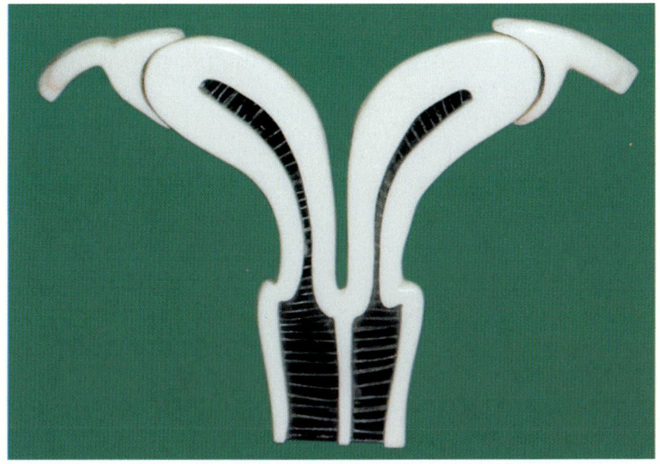

Fig. 4.34B: Didelphys

Development of Ovary

- Oogonia develop from cells of yolk sac.
- Follicular cells are derived from the celomic epithelium.
- Connective tissue stroma is from the mesoderm of the gonadal ridge medial to mesonephric ducts.

The ovaries develop in the lumbar region and later descend into pelvis.

Development of Testis (Figs 4.35A and B)

- Spermatogonia develop from cells of yolk sac.
- Basement membrane of the seminiferous tubules and Sertoli cells develop from celomic epithelium.

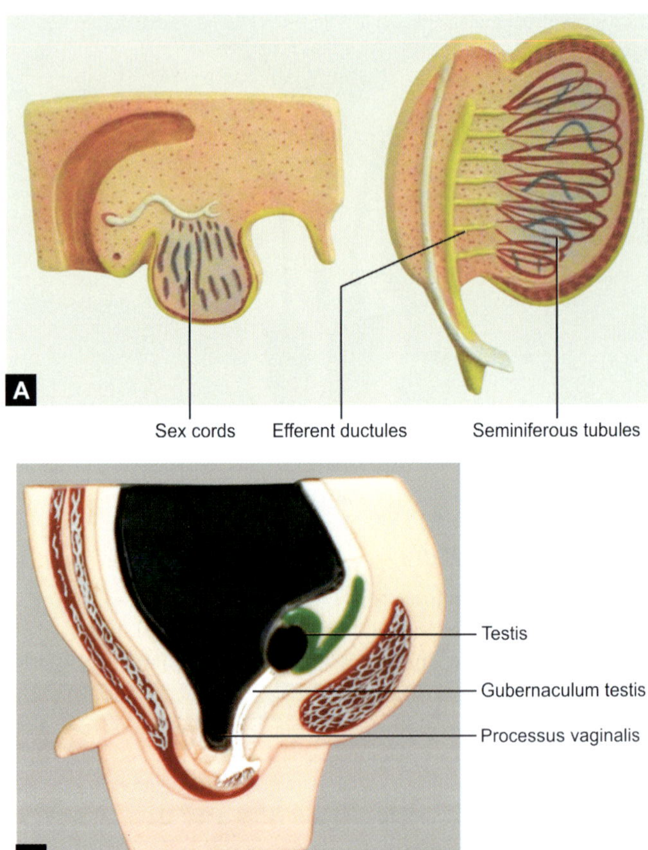

Figs 4.35A and B: (A) Development of testis; (B) Development and descent of testis

- Interstitial cells and connective tissue develop from mesoderm of gonadal ridge medial to mesonephric ducts.
- Efferent ductules are derived from the proximal 12 to 15 mesonephric tubules.
- Mesonephric ducts give rise to canals of epididymis and vas deferens.
- Cranial end of paramesonephric duct (Müllerian duct) remains as appendix of testis.
- Cranial end of mesonephric duct (Wolffian duct) remains as appendix of epididymis.

Descent of Testis

The testes descend into the scrotal sacs after the development. As the testes are descending, peritoneal covering is pulled along the testis which is called processus vaginalis.

Factors responsible for the descent:

- Gubernaculum testis is guiding force for the descent
- Increased intra-abdominal pressure facilitates the descent
- Scrotal temperature is 4° less than the intra-abdominal temperature which favors spermatogenesis
- Contraction of the arched fibers of internal oblique muscle
- Differential growth of body wall
- Testicular hormones secreted by interstitial cells of fetal testes.

Age and position of testis during its descent:

4th month of IUL: Iliac fossa.

7th to 8th month of IUL: Traverses inguinal canal deep inguinal ring.

9th month of IUL or shortly after birth reaches the scrotum.

Anomalies associated with descent of testis (Figs 4.31 to 4.41)

- *Complete congenital hernia with hydrocele:* It occurs due to complete nonobliteration of processus vaginalis. Coils of intestine descend into the scrotal sac and cause swelling which disappears when the child lies down.
- *Congenital incomplete hernia:* It occurs due to partial obliteration of processus vaginalis in which the coils of intestine do not descend into the scrotal sac.
- *Infantile hydrocele:* Nonobliteration of processus vaginalis in the distal part (in the scrotal sac).
- *Encysted hydrocele of the spermatic cord:* The processus vaginalis is closed at the deep inguinal ring and at upper part of the testis but it remains patent in the middle.

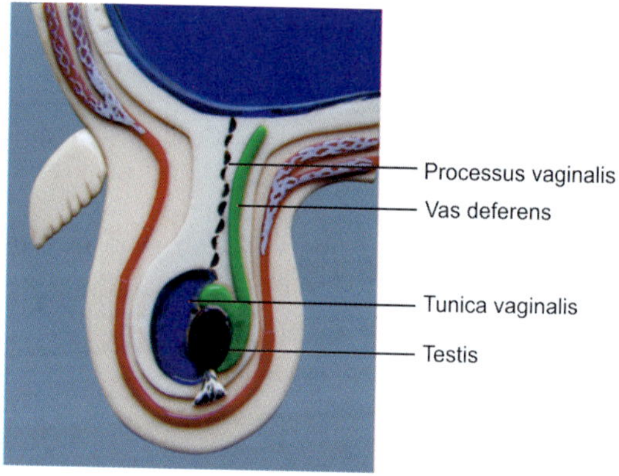

Fig. 4.36: Normal descent of testis

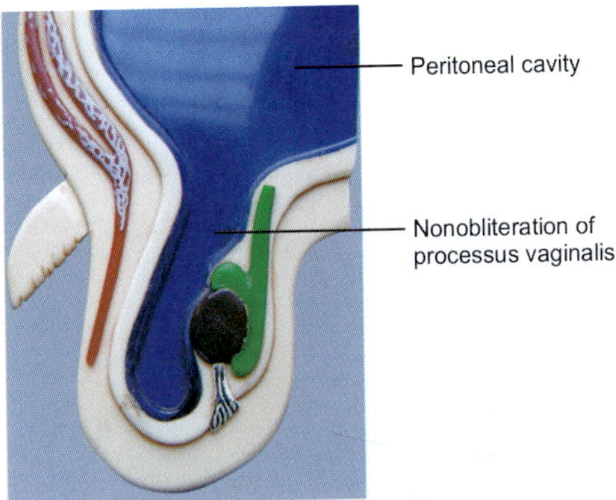

Fig. 4.37: Congenital hydrocele

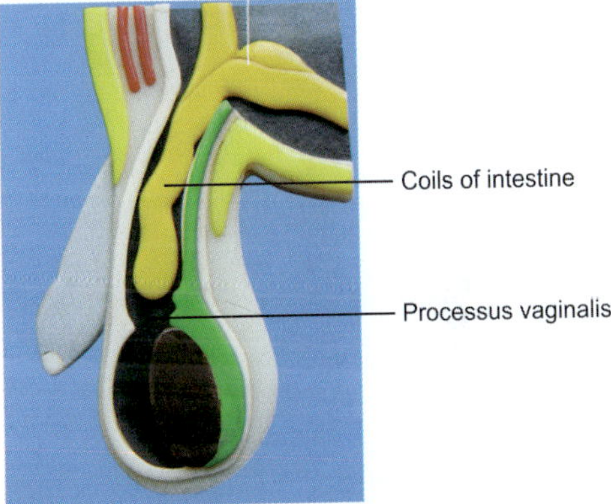

Fig. 4.38: Complete congenital hernia with hydrocele

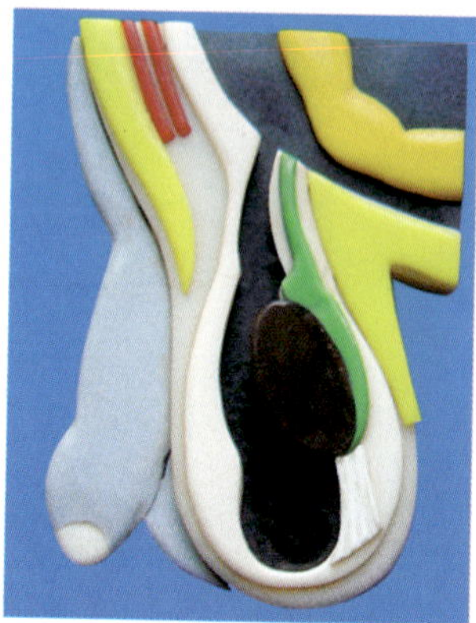

Fig. 4.39: Incomplete hernia with hydrocele

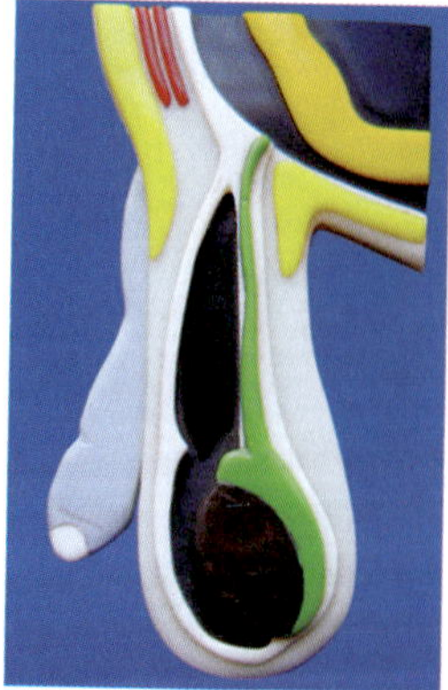

Fig. 4.40: Infantile hydrocele

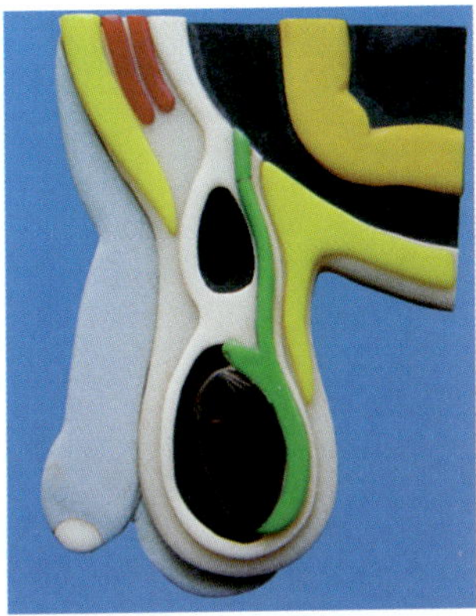

Fig. 4.41: Encysted hydrocele of the spermatic cord

DEVELOPMENT OF EXTERNAL GENITALIA (FIGS 4.42 TO 4.44)

Male

Development of External Genital Organs

The caudoventral part of urogenital sinus is dilated and is called phallic part of urogenital sinus (UGS). It is limited by urogenital membrane which is made up of external ectoderm and internal endoderm **(Fig. 4.42)**. The urogenital membrane is bounded by an elevated margin on either sides called genital folds. The genital folds are due to growth and migration of mesoderm at the caudal end of primitive streak. Another elevation called genital tubercle

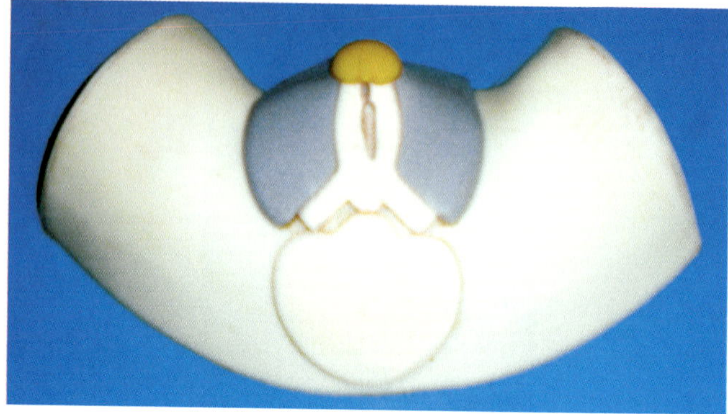

Fig. 4.42: External genitalia

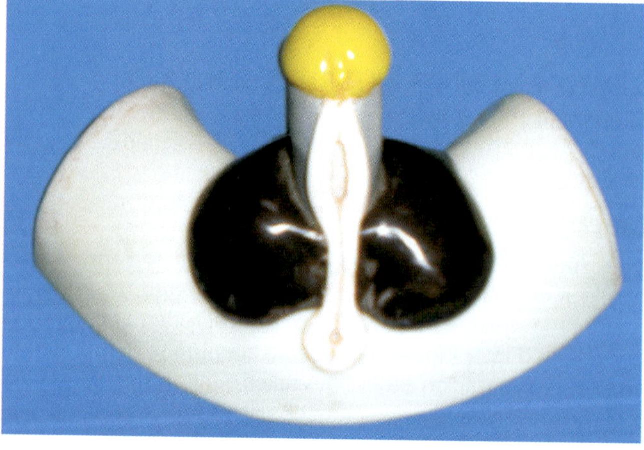

Fig. 4.43: External genitalia—male

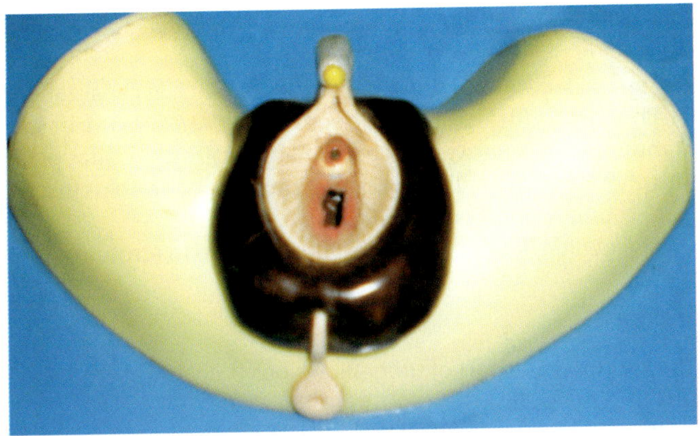

Fig. 4.44: External genitalia—female

develops in the median plane, at the cranial part of urogenital membrane. The two genital folds converge at the genital tubercle. Lateral to genital folds there are another folds called labioscrotal folds or genital swellings.

The genital tubercle elongates and forms phallus.

The phallus is invested by ectoderm and consists of mesodermal core.

As phallus elongates to form penis, a longitudinal ectodermal groove called primary urethral groove develops along caudal surface of phallus. This ectoderm is continuous with ectodermal cloaca posteriorly. This ectodermal groove extends along the inferior surface of glans penis. A sagittal solid urethral plate of endoderm arises from phallic part of urogenital sinus. It extends forwards within phallus along the primary urethral groove up to the base of glans penis **(Fig. 4.43)**.

The solid urethral plate within phallus undergoes canalization.

The urogenital membrane ruptures and establishes communication between ectodermal canalized urethral plate and endoderm of phallic part of urethra. The genital folds fuse to form spongy part of penis.

The genital (labioscrotal) swellings give rise to scrotum.

Female

The caudoventral part of urogenital sinus is dilated and is called phallic part of UGS.

It is limited by urogenital membrane which is made up of external ectoderm and internal endoderm.

The urogenital membrane is bounded by an elevated margin on either side called genital fold. The genital folds are due to growth and migration of mesoderm at the caudal end of primitive streak. Another elevation called genital tubercle develops in the median plane, at the cranial part of urogenital membrane. The two genital folds converge at the genital tubercle. Lateral to genital folds there are another pair of folds called labioscrotal folds or genital swellings **(Figs 4.42 to 4.44)**.

The genital tubercle in the midline, caudoventrally gives rise to clitoris.

The genital folds become labia minora.

The genital swellings (labioscrotal swellings) become the labia majora.

To summarize, the genital tubercle forms glans penis in male and clitoris in females.

Urethral folds form spongy urethra in male and labia minora in females.

Labioscrotal swellings form scrotum in male and labia majora in female.

DEVELOPMENT OF NERVOUS SYSTEM AND SPECIAL SENSES

Neurulation

Neural plate: Cells of the epiblast overlaying the notochord become tall columnar, producing a thickened neural plate (in contrast to surrounding epiblast that produces epidermis of skin), by the 17th to 20th day of gestation. Thus, the epiblast, finally differentiates into neuroectoderm and surface ectoderm.

Neural groove: The neural plate is transformed into a neural groove by the 23rd day. The margins of the neural groove form the neural folds.

Neural tube: The dorsal margins of the neural groove merge medially, forming a neural tube composed of columnar neuroepithelial cells surrounding a **neural cavity** on the 25th day. The neural tube has two openings—a cranial opening called anterior neuropore and a caudal opening called the posterior neuropore. The closure of the neural tube is dependent upon protein bridges bound together by calcium.

The neural tube presents three flexures—craniocaudally —cephalic flexure, pontine flexure and cervical flexure. These flexures appear due to the developing brain vesicles. There are three primary brain vesicles—prosencephalon (forebrain vesicle), mesencephalon (mid brain vesicle) and rhombencephalon (hind brain vesicle). After the formation of flexures the primary brain vesicles further differentiate into secondary brain vesicles **(Fig. 4.46A)**.

The prosencephalon differentiates into telencephalon and diencephalon.

Mesencephalon remains as mesencephalon.

Rhombencephalon differentiates into metencephalon and myelencephalon.

Table 4.3 listing the primary and secondary brain vesicles and their derivatives.

Table 4.3: Primary and secondary vesicles and their derivatives

Primary brain vesicle	*Secondary brain vesicle*	*Derivative of the wall*	*Cavity forms*
Prosencephalon	Telencephalon	Cerebral hemisphere	Lateral ventricle
	Diencephalon	Diencephalon	Third ventricle
Mesencephalon	Mesencephalon	Mid brain	Cerebral aqueduct
Rhombencephalon	Metencephalon	Pons and Cerebellum	Fourth ventricle
	Myelencephalon	Medulla oblongata	

In the process of separating from overlaying ectoderm, some neural plate cells become detached from the tube and collect bilateral to it, forming **neural crest**.

By 25 to 26 day the anterior neuropore closes and by 27th day the posterior neuropore closes.

Once closure is affected, the neural crest also begins to form.

The neural crest is the source of neurons for the peripheral nervous system as well as for chromaffin cells in the inner part (medulla) of the adrenal gland. Chromaffin cells are responsible for synthesizing and secreting two important hormones, instrumental in emotional arousal—epinephrine and norepinepherine.

Derivatives of Neural Tube and Neural Crest (Flow chart 4.1)

- Neural tube will form brain and spinal cord
- Neural crest derivatives are:
  - Ganglia of dorsal root
  - Sensory ganglia of V, VII, VIII, IX, X cranial nerves
  - Autonomic ganglia
- Enamel of tooth
- Melanocytes
- Adrenal medulla
- Parafollicular cells of thyroid
- Schwann cells
- Pia and arachnoid cells.

HISTOGENESIS OF NERVOUS TISSUE

At first, the wall of the neural tube is composed of a single layer of columnar ectodermal cells. Soon, the side-walls become thickened,

Flow chart 4.1: Derivatives of neural tube

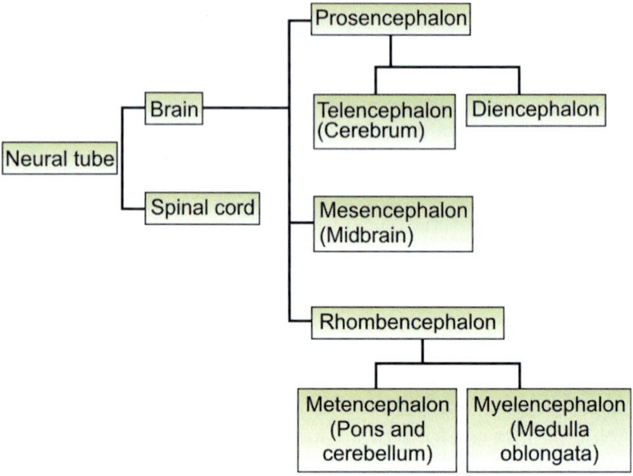

- Neural tube
  - Brain
    - Prosencephalon
      - Telencephalon (Cerebrum)
      - Diencephalon
    - Mesencephalon (Midbrain)
    - Rhombencephalon
      - Metencephalon (Pons and cerebellum)
      - Myelencephalon (Medulla oblongata)
  - Spinal cord

while the dorsal and ventral parts remain thin, and are named the **roof**- and **floor-plates (Fig. 4.45)**.

- A transverse section of the tube at this stage presents an oval outline, while its lumen has the appearance of a slit. The cells which constitute the wall of the tube proliferate rapidly, lose their cell-boundaries and form a syncytium. This syncytium consists at first of dense protoplasm with closely packed nuclei, but later it opens out and forms a looser meshwork with the cellular strands arranged in a radiating manner from the central canal.

Three layers may now be defined—an internal or ependymal, an intermediate or mantle, and an external or marginal.

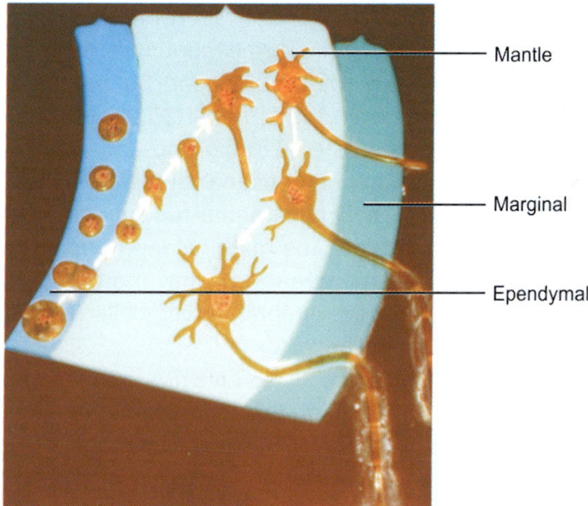

Fig. 4.45: Histogenesis of nervous system

- The **ependymal layer** is ultimately converted into the ependyma of the central canal; the processes of its cells pass outward toward the periphery of the medulla spinalis.
- The **marginal layer** is devoid of nuclei, and later forms the supporting framework for the white funiculi of the medulla spinalis.
- The **mantle layer** represents the whole of the future gray columns of the medulla spinalis; in it the cells are differentiated into two sets:
    a. **Spongioblasts** or **young neuroglia cells**, and
    b. **Germinal cells**, which are the parents of the **neuroblasts** or **young nerve cells**.

The triple layered arrangement seen in the spinal cord (ependymal, mantle and marginal layers) continues into the caudal portion of the brain (the medulla oblongata), but becomes more complex cranially. In the brain, the neuronal cell bodies comprising the gray matter become clustered into groups called **nuclei**.

In some parts of the brain, neurons and neuroglia differentiating from the mantle layer of the original neural tube migrate outwards through the white matter of the marginal layer and they form a peripheral, multilayered covering of gray matter. This outer covering of gray matter on the cerebral and cerebellar hemispheres is the cerebral and cerebellar cortex.

By subdivision, the germinal cells give rise to the neuroblasts or young nerve cells, which migrate outward from the sides of the central canal into the mantle layer and neural crest, and at the same time become pear-shaped; the tapering part of the cell undergoes still further elongation, and forms the axis cylinder of the cell.

Neurons develop from neuroblasts of the neuroepithelium and migrate into the mantle layer. The neuroblast develops into a bipolar cell having a primitive axon and dendrite. The single dendrite degenerates and is replaced by multiple dendrites forming a multipolar neuroblast. Axons have few branches and their axonal growth cones are directed to their targets by tropic factors. Primitive supporting cells are glioblasts and migrate into the mantle and marginal layers to become astrocytes and oligodendrocytes. Microglia are derived from invading blood monocytes.

- Many neuroblasts migrate (travel) like ameba, that is, by extending a part of itself, grabbing something to hang onto and then pulling the rest of the cell along. In the neocortex and cerebellum, neuroblasts must travel to their final destinations, locating themselves in the correct cell layer, orient themselves, and initiate dendritic growth to make the appropriate synaptic connections.

- The lateral walls of the medulla spinalis continue to increase in thickness, and the canal widens out near its dorsal extremity, and assumes a somewhat lozenge shaped appearance. The widest part of the canal serves to subdivide the lateral wall of the neural tube into a **dorsal** or **alar**, and a **ventral** or **basal lamina**.
- In the 3rd and 4th week the development of the spinal cord is seen, and by the end of the 4th week of gestation, the marginal layer nerve fibers appear and begin to accept fibers of ganglion, nerves that are sent into them from the peripheral ganglia. Once connected, they begin to function.
- *Fifth week:* The cerebral hemispheres differentiate.
- By the end of the 6th week the rudimentary development of the five brain vesicles is complete. The cerebral hemispheres have grown and now cover the diencephalon, the mesencephalon and the cerebellum. As these two hemispheres grow toward each other, they meet in the middle and continue their growth downwards. The membrane that separates them is the falx cerebri—a part of the dural membrane system of the meninges, of which it is the outer layer—the dura mater. The fissure thus created is known as the longitudinal fissure **(Figs 4.46 to 4.51)**.
- By the 7th week of life, the pineal gland, choroid plexus are formed from the roof of the diencephalon and their specialized cells secrete cerebrospinal fluid.

Rhombic lip gives rise to cerebellum.

The part of the neural tube which is forming pons and open part of medulla oblongata consists of laterally placed alar laminae which are separated from the basal laminae by sulcus limitans.

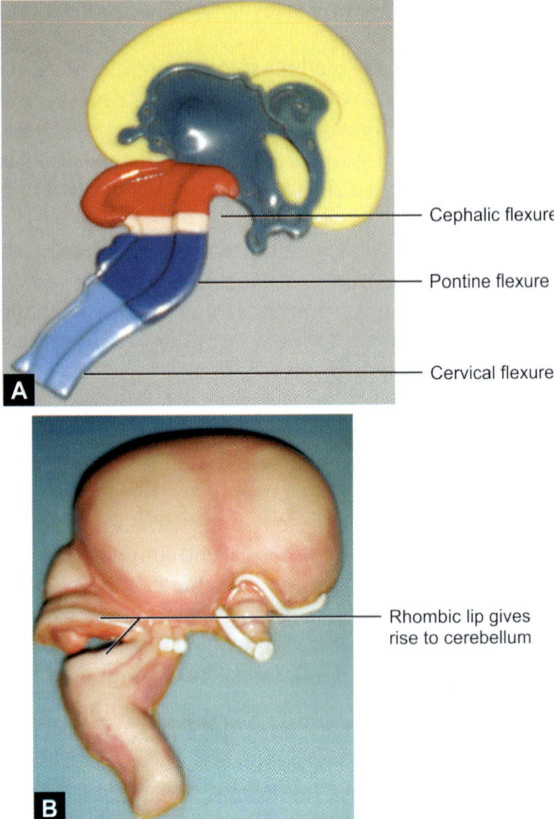

Figs 4.46A and B: (A) Brain vesicles showing flexures; (B) Rhombic lip from metencephalon gives rise to cerebellum

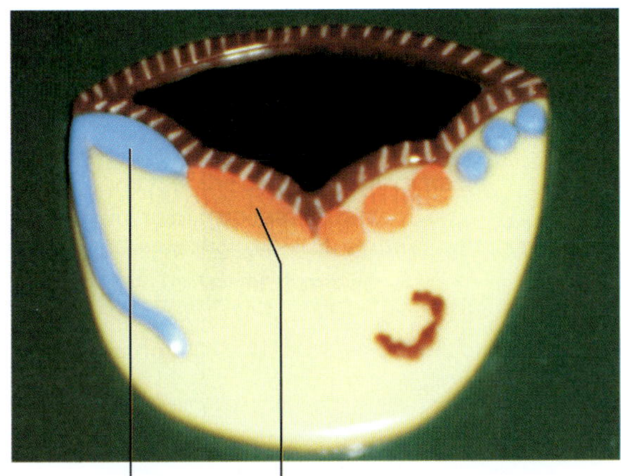

Fig. 4.47: Open part of medulla oblongata

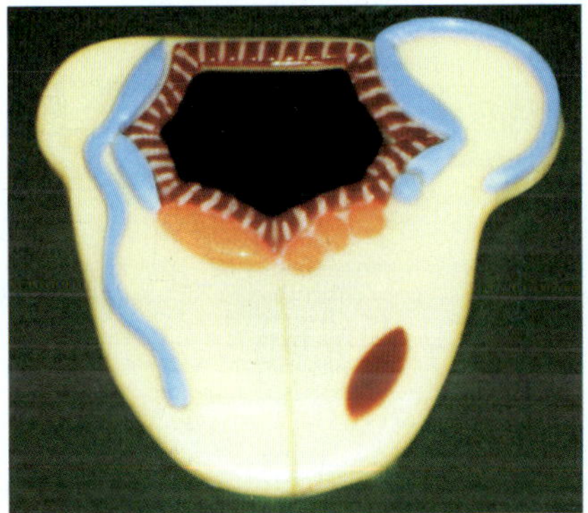

Fig. 4.48: Pons

Fig. 4.49: Midbrain

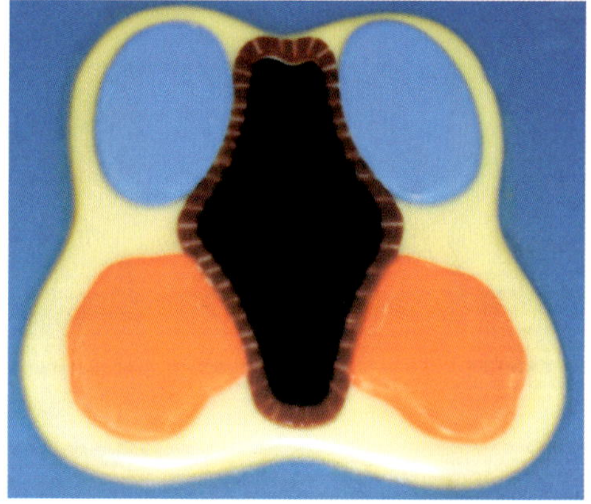

Fig. 4.50: Development of spinal cord

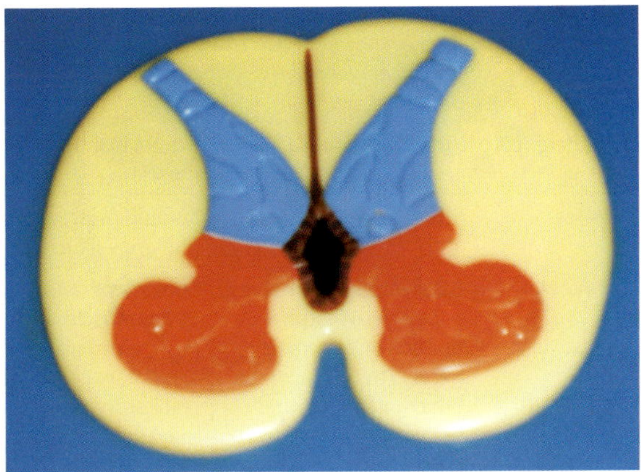

Fig. 4.51: Spinal cord

Each basal lamina contains somatic and visceral motor neurons. Each alar lamia contains association neurons which synapse with afferent fibers from sensory neurons of dorsal root ganglia.

By 28th day, all motor nuclei of cranial nerves are distinguishable. Each basal lamina divides into three columns and each alar lamina divides into four columns in the brainstem.

They are (from medial to lateral) given further.

BASAL LAMINA (TABLE 4.4)

- Somatic efferent column
- Special visceral efferent column
- General visceral efferent column

Table 4.4: Classification and location of basal lamina		
Columns	*Cranial nerves of brainstem*	*Area of supply*
Somatic efferent	3rd, 4th, 6th and 12th	Extraocular and tongue muscles
Special visceral efferent	5th, 7th, 9th, 10th and 11th	Striated muscles of pharyngeal arches
General visceral efferent	Parasympathetic nerves —Edinger-Westphal nucleus of 3rd cranial nerve, 7th cranial nerve, 9th cranial nerve and dorsal nucleus of 10th cranial nerve	Sphincter pupillae, ciliaris muscle, smooth muscle and glands of alimentary canal and respiratory system, salivary glands, cardiovascular system, pelvic organs

ALAR LAMINA (TABLE 4.5)

- General visceral afferent column
- Special visceral afferent column
- General somatic afferent column
- Special somatic afferent column

Table 4.5: Classification and location of alar lamina		
Column	*Cranial nerves associated with brainstem*	*Area of sensation*
General visceral afferent	9th and 10th cranial nerves	Sensory receptors from thorax, abdomen and pelvic viscera

Contd...

Contd...

Column	*Cranial nerves associated with brainstem*	*Area of sensation*
Special visceral afferent	Nucleus tractus solitarius, Salivatory nucleus, 7th, 9th and 10th	Tongue—Taste
General somatic afferent	5th, 7th, 9th and 10th cranial nerves	Skin of head and neck, mucosa of oral cavity, nasal cavity, pharynx and larynx
Special somatic afferent	8th cranial nerve	Vestibule and cochlea (special sense of balance and hearing)

In medulla oblongata, the inferior olivary nucleus is from alar lamina.

In pons the alar lamina migrates ventrally to form pontine nuclei.

In midbrain, alar lamina gives rise to collicular nuclei and substantial nigra. Basal lamina gives rise to red nuclei, 3rd and 4th cranial nerve nuclei.

In spinal cord, basal lamina gives rise to anterior gray column (motor) and alar lamina gives rise to posterior gray column (sensory).

The lateral part of basal lamina in thoracolumbar and sacral regions give rise to sympathetic and parasympathetic gray column (special visceral efferent column) respectively.

SPECIAL SENSES

Formation of the Eye

- Both eyes are derived from a single field of the neural plate. The single field separates into bilateral fields associated with the diencephalon. The following events produce each eye **(Figs 4.52 to 4.56)**:
  - A lateral diverticulum from the diencephalon forms an *optic vesicle* attached to the diencephalon by an *optic stalk.*
  - A *lens placode* develops in the surface ectoderm where it is contacted by the optic vesicle.

    The lens placode induces the optic vesicle to invaginate and form an **optic cup** while the placode invaginates to form a *lens vesicle* that invades the concavity of the optic cup.
  - An *optic fissure* is formed by invagination of the ventral surface of the optic cup and optic stalk. Hyaloid *artery, a branch of central artery of retina,* invades the fissure to reach the lens vesicle.

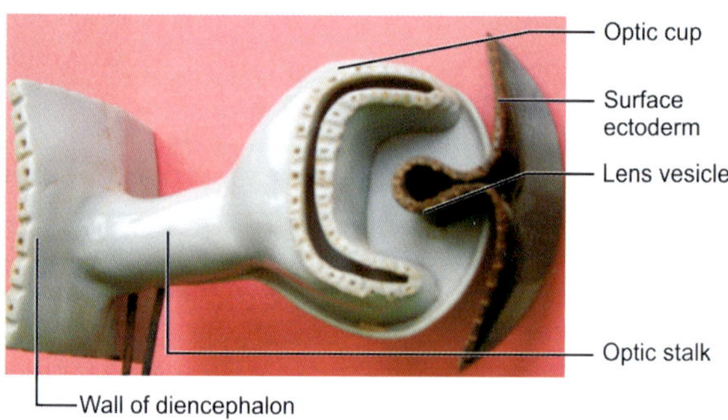

Fig. 4.52: Development of eyeball

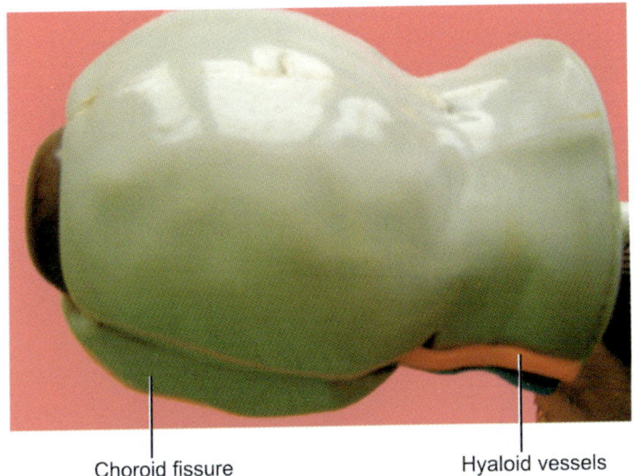

Fig. 4.53: Choroid fissure

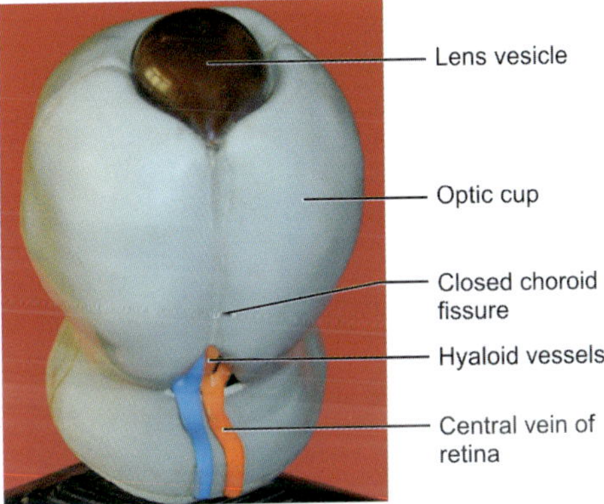

Fig. 4.54: Choroid fissure and hyaloid artery

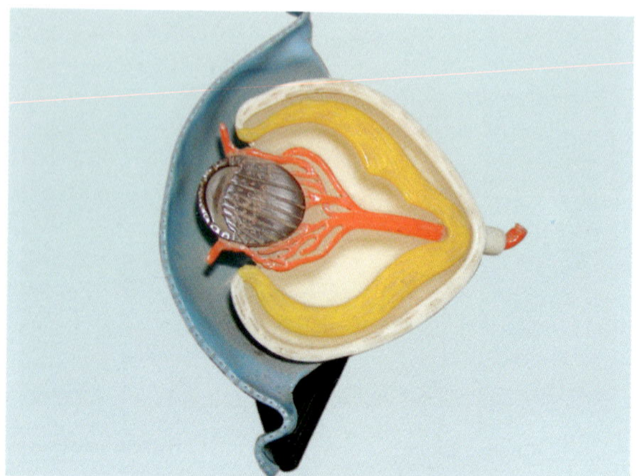

Fig. 4.55: Optic cup and hyaloid artery

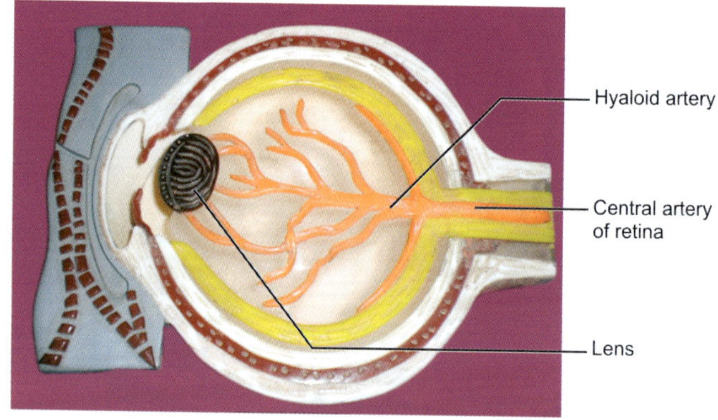

Fig. 4.56: Layers of the fully developed eyeball

- The optic cup forms the *retina* and contributes to formation of the *ciliary body* and *iris*. The outer wall of the optic cup forms the *outerpigmented layer of the retina*, and the inner wall forms *neural layers of the retina*.
- The optic stalk becomes the optic nerve as it fills with axons traveling from the retina to the brain.
- The lens vesicle develops into the lens, consisting of layers of lens fibers enclosed within an elastic capsule.
- The vitreous compartment develops from the concavity of the optic cup, and the *vitreous body* is formed from ectomesenchyme that enters the compartment through the optic fissure.
- Ectomesenchyme (from neural crest) surrounding the optic cup condenses to form inner and outer layers, the *future choroid and sclera*, respectively.
- The *ciliary body* is formed by thickening of choroid ectomesenchyme plus two layers of epithelium derived from the underlying optic cup; the ectomesenchyme forms *ciliary muscle and the collagenous zonular fibers* that connect the ciliary body to the lens.
- The *iris* is formed by choroid ectomesenchyme and the superficial edge of the optic cup.
- The outer layer of the cup forms *dilator and constrictor muscles* and the inner layer forms pigmented epithelium.
- The ectomesenchyme of the iris forms a *pupillary membrane* that conveys an anterior blood supply to the developing lens; when the membrane degenerates following development of the lens, a pupil is formed.
- The *cornea* develops from two sources:
  1. The layer of ectomesenchyme that forms sclera is induced by the lens to become inner epithelium and stroma of the cornea.

2. While surface ectoderm forms the outer epithelium of the cornea.

- The *anterior chamber* of the eye develops as a cleft in the ectomesenchyme situated between the cornea and the lens.
- The *eyelids* are formed by upper and lower folds of ectoderm, each fold includes a mesenchyme core; the folds adhere to one another but they ultimately separate either prenatally (ungulates) or approximately 2 weeks postnatally (carnivores).
- Ectoderm lining the inner surfaces of the folds becomes *conjunctiva, and lacrimal glands* develop by budding of conjunctival ectoderm.
- *Skeletal muscles* that move the eye (extraocular eye muscles) are derived from rostral somitomeres (Innervated by cranial nerves III, IV, and VI).

Formation of the Ear (Figs 4.57A and B)

- The ear has three components: external ear, middle ear, and inner ear.
- The inner ear contains:
  1. Sense organs for hearing (cochlea) and
  2. Detecting head acceleration (vestibular apparatus), the latter is important in balance.
- The middle ear contains bones (ossicles) that convey vibrations from the tympanic membrane (ear drum) to the inner ear.
- The outer ear.

Inner Ear

- An otic placode develops in surface ectoderm adjacent to the hindbrain in the floor of the first pharyngeal cleft.

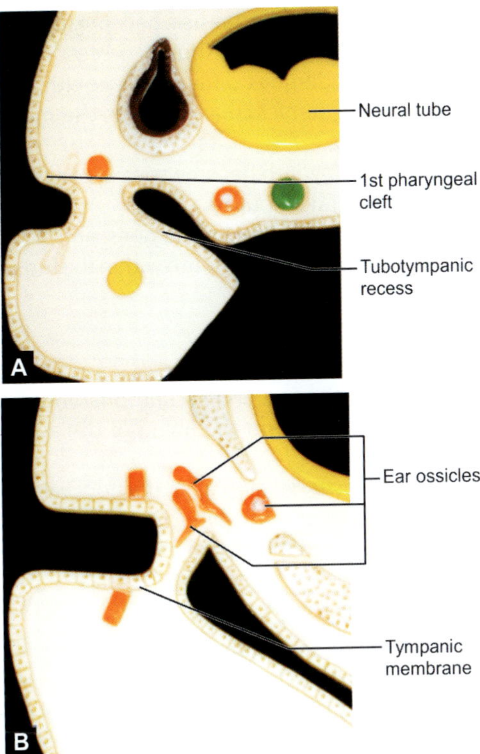

Figs 4.57A and B: Development of ear

- The placode invaginates to form a cup which then closes and separates from the ectoderm, forming an otic vesicle (otocyst).
- Otic capsule, composed of cartilage, surrounds the otocyst.

- Some cells of the placode and vesicle become neuroblasts and form afferent neurons of the vestibulocochlear nerve (VIII).
- The otic vesicle undergoes differential growth to form the cochlear duct and semicircular ducts of the membranous labyrinth.
- Some cells of the labyrinth become specialized receptor cells found in maculae and ampullae.
- The cartilaginous otic capsule undergoes similar differential growth to form the osseous labyrinth within the future petrous part of the temporal bone.

Middle Ear

- The dorsal part of the first and second pharyngeal pouches (tubotympanic recess) form the lining of the auditory tube and tympanic cavity.
- The *malleus and incus* develop as endochondral bones from ectomesenchyme in the *first branchial arch* and the *stapes* develops similarly from the *second arch*.

Outer Ear

- The tympanic membrane is formed by apposition of endoderm and ectoderm where the first pharyngeal pouch is apposed to the cleft between the first and second branchial arches (the first pharyngeal membrane).
- The external auditory meatus is formed by the groove/cleft between the first and second branchial arches. The arches on either side of the first cleft expand in the form of auditory hillocks laterally to form the wall of the canal and the auricle (pinna) of the external ear.

The Taste

Taste Buds

- The taste buds become apparent during the 8th week of gestation, and by the 14th week the taste sensation is formed. At birth, infants express positive and aversive facial responses to tastes.
- Taste buds are groups of specialized (chemoreceptive) epithelial cells localized on papillae of the tongue. Afferent innervation is necessary to induce taste bud formation and maintain taste buds. Cranial nerves VII (for rostral two-thirds of tongue) and IX (for caudal third of tongue and circumvallate papillae) innervate the taste buds of the tongue.

Olfaction (The Sense of Smell)

- Olfaction (smell) involves olfactory mucosa located caudally in the nasal cavity and the vomeronasal organ located rostrally on the floor of the nasal cavity. Olfactory neurons are chemoreceptive; their axons form olfactory nerves (I).
- An olfactory (nasal) placode appears bilaterally as an ectodermal thickening at the rostral end of the future upper jaw; the placode invaginates to form a nasal pit that develops into a nasal cavity as the surrounding tissue grows outward; in the caudal part of the cavity, some epithelial cells differentiate into olfactory neurons.
- The vomeronasal organ develops as an outgrowth of nasal epithelium that forms a blind tube; some epithelial cells of the tube differentiate into chemoreceptive neurons.

Sensory Development—Touch

The somatosensory system begins to develop during gestation. The nervous system, which is the message carrier to the brain for the senses, begins to develop at the 3rd week of gestation. At the 9th week of gestation the sensory nerves have developed and are touching the skin. By the 22nd week of gestation, the fetus is sensitive to touch and temperature. At birth, the sense of touch can be observed through the infant's reflexes when it comes in contact with different stimuli.

DEVELOPMENT OF ENDOCRINE GLANDS

Development of Pituitary Gland

It develops from two sources. They are **(Figs 4.58A and B)**:

1. Ectodermal roof of stomodeum.
2. Neuroectodermal diverticulum from the floor of the third ventricle (hypothalamus).

The ectodermal roof of the stomodeum grows cranially towards the base of the diencephalon as a diverticulum. It is called Rathke's pouch.

The pouch becomes elongated and gets separated from the ectodermal stomodeum to form vesicle. This vesicle approaches the diverticulum from the floor of the diencephalon and surrounds it.

The anterior wall of the vesicle grows more and becomes pars distalis, the tubular prolongation of the vesicle forms pars tuberalis and the posterior wall of the Rathke's vesicle becomes the pars intermedia. The cavity of the Rathke's' vesicle becomes intermediate cleft.

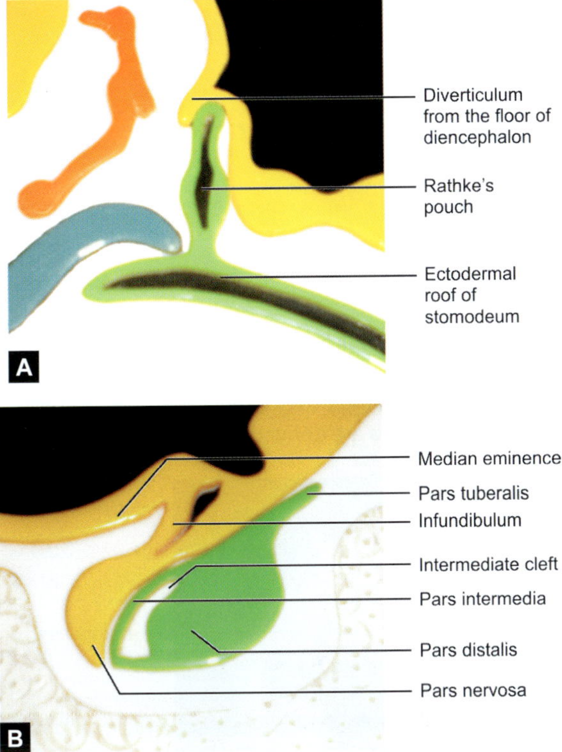

Figs 4.58A and B: Development of pituitary—Parts of pituitary gland. Green color showing surface ectoderm and yellow color showing neuroectoderm

The down growth from the diencephalon retains its continuity and becomes pars nervosa, pars infundibularis and median eminence which is part of hypothalamus.

Development of Suprarenal Glands

The gland consists of:
- Outer cortex and
- Inner medulla.

Development of suprarenal gland is as follows **(Figs 4.59A to C)**:
- The adrenal gland begins to develop in the 5th week of IUL.
- The cells of **cortex** arise from **celomic epithelium** that lies in the angle between the developing gonad and mesentery.
- The cells of medulla are derived from **neural crest**.

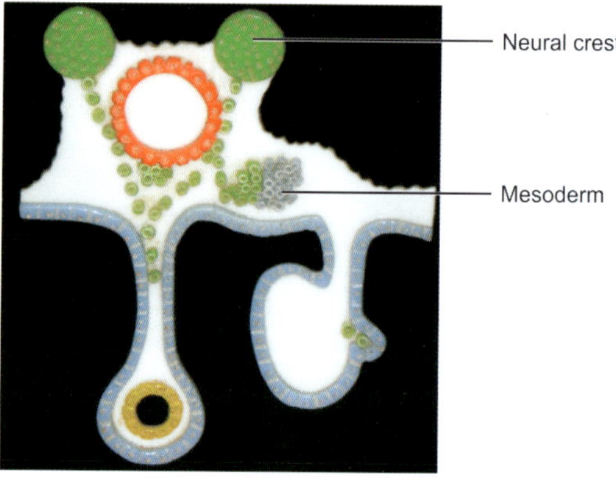

Fig. 4.59A

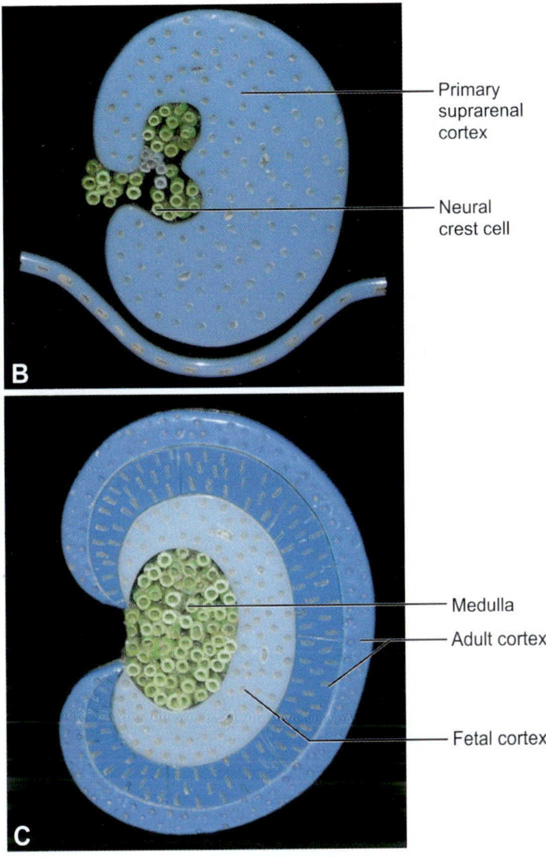

Figs 4.59B and C

Figs 4.59A to C: Development of suprarenal gland

Development of Cortex

The cells that are formed first are large and acidophilic and surround the medulla. They form fetal cortex. The cortex engulfs the medulla until the medulla is completely encapsulated. The fetal cortex disappears after birth.

The celomic epithelium gives rise to small cells that surround the fetal cortex. This is the definitive cortex.

Development of Medulla

The cells of medulla are derived from neural crest.

These cells migrate to the region of developing cortical cells and come to be surrounded by them.

- Throughout the remaining fetal period, two cortical layers surround the gland, first the zona glomerulosa and then the zona fasciculata.
- A final layer, the zona reticularis forms after birth at about 3 years.
- The fetal adrenal glands are relatively large.
- At 4 months of intrauterine life, the glands are larger than the fetal kidney.
- At birth they are relatively large (about 20 times their relative size in the adult).
- After birth, the fetal cortex rapidly undergoes involution (regression).
- It is the inner "fetal zone" of the adrenal cortex that persists during gestation, suggesting that these cells are involved in a specific function during pregnancy.

Development of Thyroid Gland

It develops as a midline endodermal diverticulum from the floor of the developing foregut at the developing tongue region. It is called the midline thyroid diverticulum. It grows up to the level of 4th pharyngeal pouches. The elongated stalk is called the thyroglossal duct and the caudal end of the duct there is thyroid diverticulum. This grows into lobes and the isthmus. The lobes are reinforced by the ventral parts of the fourth pouches called the lateral thyroid divericula. The parafollicular cells of thyroid gland develop from ultimobranchial bodies which contain the incorporated neural crest cells.

The connective tissue and vessels develop from the visceral mesoderm.

Development of Parathyroid Glands

They are four in number. On either side of the midline. Two are superior and two are inferior.

Superior parathyroids develop from endoderm of the dorsal part of fourth pouches and inferior parathyroids are from the endoderm of the dorsal part of third pouches.

The dorsal part of the third pouch migrates inferior to fourth pouches along with the migrating thymic lobes which bring them below the level of the fourth pouches.

The connective tissue and the vessels develop from the surrounding visceral mesoderm.

Index

Page numbers followed by *f* refer to figure and *t* refer to table.

D

E

F

G

H

I

K

L

M

N

O

P